Embryos under the Microscope

JANE MAIENSCHEIN

Embryos under the Microscope

THE DIVERGING MEANINGS OF LIFE

Harvard University Press

Cambridge, Massachusetts
London, England 2014

Printed in the United States of America

Library of Congress Cataloging-in-Publication Data

Maienschein, Jane.
Embryos under the microscope : the diverging meanings of life / Jane Maienschein.
pages cm
Includes bibliographical references and index.
ISBN 978-0-674-72555-3 (hardcover : alk. paper)
1. Embryology, Human—Popular works. 2. Human embryo—Popular works.
3. Developmental biology—Popular works. I. Title.
QM603.M35 2014
612.6'4—dc23 2013039073

Contents

Preface

When I have presented educational programs for federal judges under the auspices of the Federal Judicial Center or through the circuit courts, the judges have mostly started out without much science background—just as I started out without any significant political background. They mostly have not yet had to make court decisions themselves that have related to embryos, though a few have. But once they begin to hear about the issues that could arise and about the underlying biological facts, they become very interested and want to know more. They see that they are likely to have cases in the future for which understanding of what is at issue and what is at stake can help them judge wisely.

I also observed this eagerness for more information when serving as a senior congressional fellow, working in Congressman Matt Salmon's office during the United States 105th Congress in 1997 and 1998. Dolly, the cloned sheep, and then stem cell research both appeared during this time, and some of the initial responses were astonishing. One congressman wanted to outlaw cloning completely, but he had so little understanding of what is involved in the reproductive and developmental processes that his proposed bill would have accidentally prohibited any mother from having natural identical twins. Others started with sensible ideas but then confronted constituents who argued with widely divergent but deeply held beliefs and little in the way of biological facts about stem cell research. I saw many people eager to learn and eager for reliable knowledge, yet they were caught in political and scientific confusion, sometimes caused or reinforced by the media.

Individuals who want to have a baby also need solid information. Those using fertility treatments have to make decisions about what embryos are, and they report often feeling that they do not have the right kind of information to make wise choices. This can lead them to rely on partial or simplistic ideas that are often just wrong biologically and may lead to unfortunate choices.

Members of the public are in a similar situation: inquiring minds do want to know, but it is not always obvious where to get reliable knowledge. Political discussions often serve as a sort of centrifuge, throwing the rhetoric away from the middle and toward the most extreme and farthest distant views, yet I optimistically feel that most of the time most people really like to be informed about science. We need interpreters to help us sort through the seemingly insurmountable piles of things to read, to present the story of science—and that is what this book will do.

This story is told historically, though not always strictly chronologically. Why do we study history? Because history can make science more reflective as well as make us better judges of scientific ideas and their implications. An old adage states that we study history to avoid repeating it. That makes sense in some cases: there are many unfortunate historical episodes we should try to avoid going through again. In other cases, if we understand the reasoning of the past, we may want to repeat those episodes, rather than avoid them.

History shows us that some scientists in the past addressed just the right questions, in just the right way, but they were hampered by limitations in technology, or choosing the wrong organism to study, or insufficient finances, for example. In these cases, reflecting on why events took the turns they did can tell us a great deal. Scientists today may be able to go back with new tools and expanded knowledge in hand to productively pick up where those previous scientists had left off and perhaps make new progress.

Embryo research provides some especially compelling examples. By the 1950s, Robert Briggs and Thomas King had transplanted nuclei from one frog embryo into another and watched the resulting frog develop. John Gurdon then showed that in a smaller percentage of cases, he could transplant nuclei from adult frog cells into eggs and watch them develop into tadpoles. This work, the first cloning studies of animals, showed that researchers could reprogram cells through nuclear transplantation. Gurdon's results suggested that reprogramming was possible, and that adult differentiated cells (or at least their nuclei) could be changed into an undifferentiated state. But developmental biologists at the time relied on the major underlying assumption that cells develop in one direction only. They thought that once a cell is differentiated, it stays that way and cannot go backward. Gurdon's cloning example seemed like a special case, as textbooks of the time make clear that one-way differentiation was taken as given. This assumption turns out to have been wrong. It is worth revisiting this example and asking why researchers made the assumptions they did and what caused the field to change its collective mind.

In fact, those studying development had two deeply held assumptions before the late 1990s. First, they believed that differentiation works in one direction: once a cell becomes differentiated as a particular kind of cell, it will remain that kind. Second, they thought that the farther a developmental stage progresses, the more fixed and inflexible the developmental process becomes. These assumptions guided the thinking about embryos until 1997. Then we heard about the cloning of Dolly the sheep. The year after, we learned about the culture of human embryonic stem cells. These events provoked a flood of research, which has changed both of the fundamental assumptions. In addition, parallel discoveries have challenged assumptions about the complex regulatory processing that guides gene expression. Furthermore, the changes in these underlying assumptions have raised the possibility of actually guiding the construction of embryos.

Historians and philosophers of science remind us of where we have been and help us to understand the underlying assumptions of scientific research at any given time, even when the researchers themselves have not articulated them. Historical and philosophical reflection can help illuminate why researchers made particular choices in the past, and that in turn can help us reflect on the choices being made today.[1]

Understanding the science alone, even in the context of historical perspective, will therefore not tell us what ethical considerations matter or what policy decisions to make. Yet knowledge about embryo research does provide a framework of reality that we are foolish to ignore. In his letter about the motion of the heavens to the Grand Duchess Christina, Galileo famously claimed that he intended his astronomical studies to tell us "not how to go to heaven, but how the heavens go." Historians will go on to point out that he did not really stop there—in fact, he made religious as well as scientific claims. Indeed, the two overlap at times in quite complex ways.

In this book, as in Galileo's case, understanding the biological facts about embryos tells us much about nature but not everything we need to know in order to think about social questions about nature and science. There is room for belief and values in making social decisions, and though the book's narrative tells the story of science, we return in the conclusion to reflections on larger implications.

Embryos under the Microscope

1

Recurring Questions, Seeing and Believing

Looking at embryos without a microscope does not show much by itself. Human embryos are too tiny to see at all other than as teensy specks in a laboratory dish at a fertility clinic. Frog embryos are large enough to see, but not with much detail: a big egg cell divides into other cells and then gives rise to a tadpole, which swims around for a while then metamorphoses through a process of changing shape into a frog. Chick embryos are inside eggshells. Other species form in similar ways, and without a microscope to magnify the cells, we cannot see much of the intricate detail that is there in any of them.

In the past, the lack of empirical observed information required us to imagine what happens at the earliest stages of developing life. Some people (historically known as *preformationists*) believed that what exists from the beginning is a miniaturized version of the adult form. Others surmised that the form is not present initially, that instead something causes it to emerge over time, perhaps some special living or vital factor. The gradual changes seen in developing frogs would be evidence for such claims among those who thought about development in this second way (known as *epigenesists*), who imagined the form as emerging over time, one stage at a time. Alternatively, some in-between idea might suggest that in the very beginning nothing is formed, yet that some predetermined information, or some special force, or even some kind of soul directs the material to take on just the right kind of form. Throughout history, philosophers advanced versions of all these different interpretations of how development occurs.

In all these cases, the embryo remained hypothetical. It was a metaphysically inspired entity, imagined based on underlying assumptions about what exists in the world and how it works. The data come from belief rather than empirically grounded knowledge. Thinkers would have little control over such a hypothetical entity because they would not understand the causes of its development. This hypothetical, metaphysical, imagined embryo has played a significant role in history, and this is the way that people thought about the animals they bred in agriculture and even about the early stages of their own gestating offspring.

Direct observation changed understanding of embryos. At first, led by Aristotle, a few philosophers inspected developing embryos without the aid of a microscope. By cutting a small hole in the shell of a chick's egg, Aristotle was able to watch form emerge from previously amorphous material. His observations emphasized an epigenetic rather than a performationist interpretation, but he could not see very much that way. Thus, the embryo still remained largely hypothetical until the introduction of the microscope in the late seventeenth century.

With the use of a microscope, better information about structure and changes in embryos came quickly. The period from the late seventeenth through the end of the nineteenth century brought accumulating knowledge and increased understanding of what embryos are and how they develop. With this information came new ideas about the meaning of life. By the end of the nineteenth century, researchers in the biological sciences had almost entirely rejected vitalist interpretations, had set aside nonscientific ideas of souls, and had come to emphasize the material basis of life.

By 1900, therefore, the embryo was a biological object for investigation. Yes, it is alive. Yes, it is the earliest stage of life. Yes, it is an individual, developing organism. But it is also a biological and material object to be studied, one that follows an internal logic and processes of its own and in response to its environment. This biological embryo became the focus of an experimental life of its own.

By 2000, it had become possible to analyze, deconstruct, reconstruct, and even construct almost from scratch the embryos of various species. Stem cell research, cloning, and synthetic biology are all methods for engineering embryos. The idea of engineering embryos had already appeared by 1900, but it took a century of research to gain the means for performing the work extensively. We are now in exciting times, with many prospects for using our knowledge of development to create new therapies and improve our quality of life.

The engineered or constructed embryo is very real—and it is also very frightening to some people. In some cases, confusion arises because so many people have never had the opportunity to examine embryos through a microscope. They have never seen the gradual emergence of new form, or the gradual appearance of the chick's beating heart. They have never seen the chromosomal actions of human cells dividing—perhaps even those of their own embryo in a fertility clinic's dish. Thus, many people cling to their metaphysically inspired, hypothetical imaginings about embryos because they simply do not know otherwise, although some do appear to find the hypothetical more comforting than the material reality.

Biological Basics

It is worth laying out some of the biological basics of what we now know about embryos before going any farther. Then the discussion will return to the scientific and public roles of embryos in historical and policy contexts. This book is about embryos generally and how our knowledge of them has evolved over time, including the divergence of understandings and meanings imputed to embryos. The embryos of different types of organisms are not the same, sometimes differing entirely, other times only in detail. For our purposes, this book will look at research on different kinds of embryos for historical understanding and then for modern times focus on the human embryo. The historical discussion illuminates how we have arrived at our present understanding of the step-by-step process of embryonic change and the complexity of each embryo and its development. And first, let us look at the biological basics for human embryos as we have come to understand them.

In humans, the process of developing from a fertilized egg into an adult follows a typical sequence of steps (though there are many ways for the process to deviate as well). A woman develops egg cells, also called *oocytes,* in her ovaries; the majority of those oocytes will die over her lifetime. The best scientific estimates suggest that human female fetuses develop with millions of oocytes in their ovaries, and that they still have nearly 1 to 2 million at birth. By the time the woman reaches puberty, when she begins to ovulate and experience monthly menstrual cycles, she has an estimated 300,000 left. Each month brings the release of one egg (or occasionally more than one) such that throughout a typical woman's life she "uses" perhaps 300 to 400 of the oocytes she was born with. Recent evidence has suggested that women may produce more oocytes during life from special stem cells that persist into her later life, but we do not presently know how many or how this occurs in any detail. Again, the majority of oocytes just die.

Men produce millions of *sperm* as well, far more than the number of oocytes that women produce. In rare cases, one sperm will reach and fertilize one egg, typically after sexual intercourse or sometimes with technological assistance. Even here, only a few eggs become successfully fertilized, and each by only one sperm cell. Then even fewer begin to undergo cell division.

At each step, there is more loss and a low probability of continued development. What is amazing is that the process actually works as often as it does. When it does work, the one fertilized egg cell divides into two cells, those two divide into four, and those four into eight. (Figure 1.1). Then the cells start dividing at different rates, until they have multiplied into around 100 cells. At this point, the cluster of cells is called a *blastocyst*. The outside is a single sphere of cells that will become the placenta, and these surround an area that is partly hollow and partly filled with the *inner cell mass* composed of many cells. This is the point when the cells inside are first called embryonic *stem cells*.

In the earliest stages of cell division, all evidence suggests that cells are just dividing. In humans, the resulting embryo does not grow larger, and the cells do not yet undergo differentiation. That means that during these earliest stages, there is no evidence of significant expression of genes translating into characteristics in the developing organism. The cells merely mechanically divide and divide.

After this point in humans, the embryo has to become implanted in a uterus to be able to continue developing. At this point also, significant gene expression begins. The model from what many consider as the golden era of genetics in the 1950s laid out what was called the central dogma, according to which the DNA (deoxyribonucleic acid) molecules in the chromosomes make smaller molecules called RNA (ribonucleic acid), which can move out and cause the production of proteins: DNA to RNA to protein. This is simple and straightforward, and although not complex enough to describe the developmental processes fully or accurately, it provides a good start for understanding the processes involved.

Gradually, developmental biologists uncovered the complexity of the processes by which genes get expressed. A great deal of regulation of the process occurs through a mix of inherited, internal, and environmental conditions that we are only just beginning to understand fully. Researchers have also added to our understanding of the

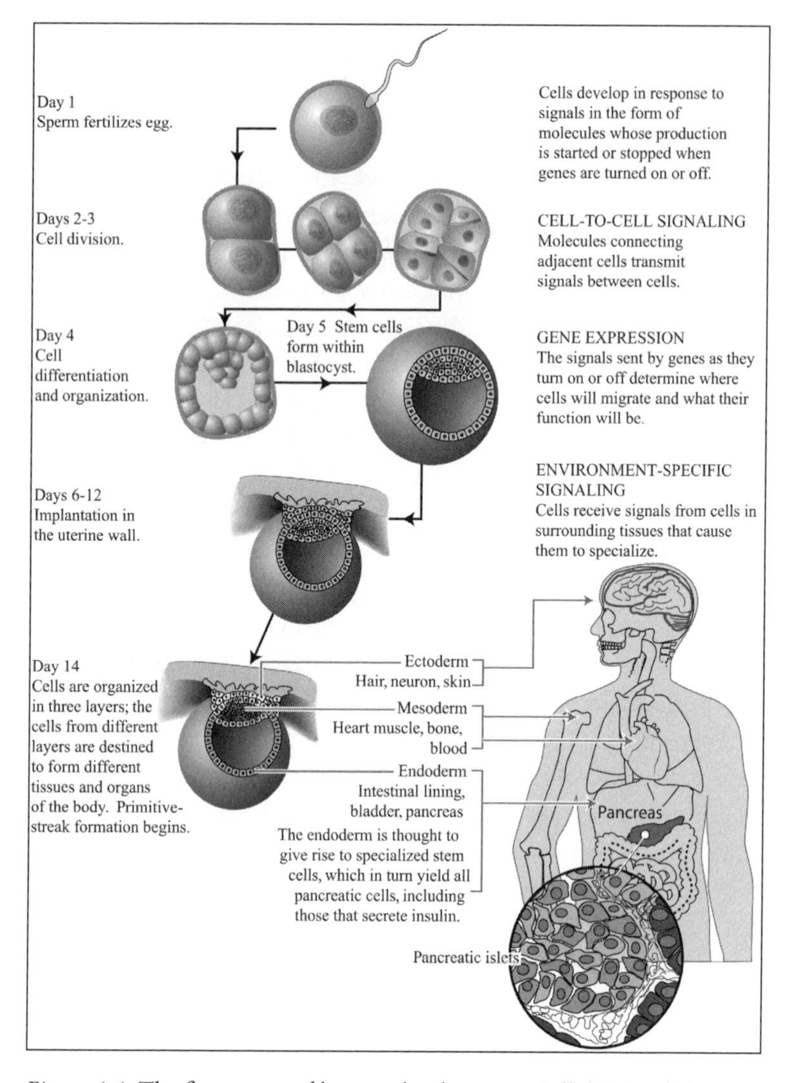

Figure 1.1. The first stages of human development. Cell division begins with fertilization of the egg cell by a sperm cell. The earliest cell divisions divide up the material into smaller units called blastomeres. Up to the eight-cell stage, each cell is totipotent, with the capacity to become a whole organism. After that, the cells begin to form a blastocyst. This consists of an outer layer of cells surrounding an inner cell mass of pluripotent stem cells. At this point, cell-cell signaling gains importance, and the blastocyst then implants in the uterus to initiate further differentiation, growth, and morphogenetic development. From Jane Maienschein, *Whose View of Life? Embryos, Cloning, and Stem Cells* (Cambridge, Mass.: Harvard University Press, 2003). Used with permission of Harvard University Press.

role of *methylation,* or the addition of extra chemical combinations called methyl groups that regulate the way gene expression occurs. The study of this area is called *epigenetics,* and it brings together what we know of genetics, development, the environment, and also the evolutionary context in which gene expression occurs.[1]

If the embryo has properly implanted, it can go on to a stage called *gastrulation,* when it begins to form what are called germ layers: the ectoderm, mesoderm, and endoderm. The germ layers will give rise to different body parts, and sorting out the precise role of each germ layer has absorbed the attention of many embryologists over the past century and a half. In addition, some researchers regard the neural crest as a separate germ layer because it plays such an important role in leading to neural structures and functions.[2]

It is no wonder that for centuries observers have been fascinated by the developmental process and have tried to explain it. Development seems to follow such predictable and regular patterns, and the processes by which the single cells come together and lead to growth, differentiation into different kinds of cells, and generation of such diverse complex organisms is itself astonishingly complex. In recent decades, researchers have shown that there is a typical path of development, but changing circumstances can prompt individuals to deviate from it in a surprising range of ways. Such changes happen naturally, but experimental embryologists can cause further deviations. Such experimentally induced results have occasionally wound up in widely circulated news stories, creating confusion because of the lack of public knowledge about embryo development.

The history of the study of embryos not only involves accumulating discoveries about the details of the developmental processes and genetic expression, but also reveals the changing underlying assumptions. Even today, when embryos are in the news and are frequently the focus of public attention, most people have never had the good fortune to study embryology or its rich history, and we have little shared public understanding of what embryos really are or how they develop. As a result, when people are confronted with confusing and conflicting claims, they have no basis for sorting through them. This lack of knowledge occurs partly because embryonic development is so complex, but also because the public presentation of scientific information is often shaped by political expediency.

This book aims to provide an accurate understanding of embryos by offering a historical look at how our knowledge has evolved in different contexts and what those changes have meant. While the social interest centers around human embryos, the story here is about understanding what embryos are and how they develop.

It has taken a lot of research on a lot of different organisms and comparisons across types of organisms to interpret human development. We know now that in humans (and many other organisms) the typical union occurs between one egg and one sperm cell. But in rare cases, two sperm cells can fertilize one egg; these eggs rarely continue to develop beyond a few cell divisions. However, in rare cases, even in humans, two eggs can join together and fuse into one developing union, which goes on to a normal healthy birth. There are also documented cases (and probably many more undocumented examples) of two fertilized eggs joining after they have started to develop. Under some conditions, these combined cells can develop into a normal whole that is a mix of two sets of genomes, called a *chimera* or a type of genetic mosaic. The earliest known cases of a mixed genotype in humans became known by accident, as the individuals looked normal in all obvious respects. Although such cases are perfectly "natural" in the sense that they occur in nature, they are neither typical or common outside the laboratory.

In nature, other ways of combining bunches of cells from early-stage embryos may occur. Still other combinations are known to be possible under laboratory conditions, or they may be only hypothetical but still imaginable. For example, try to imagine some possible combinations of cells and think about whether they would change how we should think about embryos. What if after the fertilized egg divides into two and then four and then eight cells, the embryo splits into two parts? Say that it has four cells in each of the two parts, but then each of those parts joins with a set of four cells from another embryo. (Or we could take any number of cells up to eight from each of any number of embryos.) Now we have two new chimeras that are mixes of cells from different sources, each of which was called an embryo before it divided. Are the new things also embryos? These cannot be the same embryos as before—so what has happened to that embryo that split apart? Did that original embryo—and the potential person that some people believe resides there—die? Did a new embryo begin in this case not with fertilization but with the combination of cells from different embryos? Trying to sort out such examples can become messy very quickly. What initially seemed like a nice intuitive understanding of what an embryo is does not seem clear any more. And remember that some of these cases happen in nature as well as in the laboratory.

Now let us muddy the waters even further. Instead of starting with an egg cell, try beginning with an artificial cell created in the laboratory. Say that a researcher synthesizes a set of chromosomes and puts them inside a hollow cell to make an embryo, which then develops in a way that seems perfectly normal. When does that embryo's "life" begin, and what kind of individual is it? Or say that an egg is fertilized by a sperm, but then researchers genetically engineer the resulting embryo to add or remove some of the DNA in its chromosomes. Is this still the same embryo? What if the researcher joins two fertilized eggs together—what will that result be?

Thus, we may start with some intuitive assumptions about what we mean by an embryo, but they do not all fit neatly with what happens to real embryos in the material world of experimental science and technology. Again, although researchers initially created many of these cases of cells coming together or coming apart in the laboratory, they have then discovered natural cases of the same phenomena—nature is an excellent experimentalist as well. As a result, we seem to have a muddle of what is logistically possible, in combination with sentiment and regulatory responses to recent scientific discoveries and social perception that are often based on wishful thinking about what some want to believe to be true. The result is a complex patchwork of different ideas, interpretations, and social and political responses. This book sorts through the muddle by studying historically how and why each new understanding has emerged, with the focus on the scientific understandings.

Embryos and the Public

The 2012 U.S. presidential election gave us a great example of how embryos exist in the public mind because it brought embryos into the public arena as a highly debated issue. The heat of the disputed claims and the passion of the advocates on all sides showed the extent to which embryos have taken on different meanings for different groups. Some groups hold the ancient, hypothetical view of embryos as miniature persons; others view them as a source of valuable biological material for research and medical innovation.

For example, Todd Akin, then the Missouri representative to the U.S. House of Representatives, echoed the views of a number of legislators in asserting repeatedly during his senate campaign that "legitimate" rape does not result in pregnancy. He insisted that a woman's body has a way of knowing not to become pregnant. Therefore, he maintained, we should ban all abortions after twenty weeks and make no special considerations for cases of rape. The firestorm of both protest and support for Akin revealed that he is not the only one apparently so woefully uninformed about how fertilization and conception actually work. When eggs and sperm come together in the appropriate way, the egg is fertilized, and development begins. A woman cannot will herself not to get pregnant, and there is nothing about illicit penetration through rape that causes sperm to be less effective. This hypothetical embryo is not consistent with our knowledge of the biological embryo, and Akin's underlying assumptions about pregnancy and reproduction, apparently uninformed by biological facts, could have serious policy consequences if he and others who share his views were to write them into legislation.

Another example of misguided underlying assumptions appeared in proposed legislation in the United States 112th Congress. Paul Broun (R-Georgia), Paul Ryan (R-Wisconsin, and the 2012 vice presidential candidate), and others in the House of Representatives introduced the Sanctity of Human Life Act (H.R. 212): "To provide that human life shall be deemed to begin with fertilization" (or its functional equivalent through cloning or other manipulations). They made clear that they believe that at the time of fertilization "every human being shall have all the legal and constitutional attributes and privileges of personhood."[3]

Those who identify the biological moment of fertilization as the social moment when rights are assigned to individuals are taking a huge leap of imagination about what a fertilized egg cell actually is. An individual organism does typically begin at fertilization, but it does not follow that every fertilized embryo goes on to become a formed individual organism—or even to become a person. The suggestion that a fertilized egg or embryo is legally equivalent to having personhood rests on an assumption that all biological stages should be considered legally equivalent. In reality, the majority of human fertilized eggs die in development—surely we do not want them all to be considered legal persons with full inheritance or other rights we assign to persons, which is what the proposed legislation technically would do. The facts are far more complex than this proposed legislation suggests, and the situation calls for a more nuanced understanding of what an embryo is biologically.

Fortunately, the Sanctity of Human Life Act never came to a vote in the House and therefore died—at least for the 112th Congress. Yet the fact that such a bill appeared at all, in the volatile climate of a presidential election, with a range of positive public comments about their efforts to enact it, shows the power of a socially and metaphysically inspired hypothetical idea of the embryo that does not resemble the biological embryo. Although some critics dismissed the bill's proponents as outliers and the bill itself as political grandstanding, it is important to try to understand why its advocates feel so strongly and fail so seriously in understanding the biological facts. This bill reflects a popular position based on intuition and belief; it is grounded in preexisting metaphysical assumptions about what exists in the world and about life.

These political examples of the triumph of metaphysical assumptions were balanced by other events that reflect public interest in biological advances. During the most intense period of the 2012 presidential election, the Nobel Prize for Physiology or Medicine was awarded to John Gurdon and Shinya Yamanaka "for the discovery that mature cells can be reprogrammed to become pluripotent." This celebration of genetic reprogramming, stem cell research, and its possible applications provides confirmation that knowledge about development is desirable. The Nobel committee endorsed this kind of research as biologically and medically important (and we will return to consider the discoveries in more detail later). Discussions at the time suggested that the Nobel committee's announcement was reinforcing already deeply held beliefs, so advocates for federal funding for stem cell research saw the prize as an endorsement. Those lobbying for federal funding for stem cell research cited the Nobel Prize decision as supporting their position, and opponents saw it as a reason to renew their work to prevent such funding.

At the same time, researchers were producing other tangible results in the laboratory, showing the complexities of biological development and just how far they can manipulate and change what had previously been assumed to be an already differentiated, formed, and determined organism. The announcement in 2012 from Kyoto, Japan, of the birth of live, apparently healthy mice may not seem like a major breakthrough deserving front page coverage. Yet the story merited that attention because researchers had created the eggs in the laboratory. These eggs had started as stem cells, which the researchers then cultured with the precise medium to cause them to differentiate as egg cells, and then fertilized the eggs, which resulted in live births. Researchers had been trying to turn stem cells into eggs for some time, but skeptics assumed that barriers would prevent this developmental step. Now it is likely that other researchers are working to perform the same procedure with human stem cells, because it would allow medical treatment for some infertility problems. However, this particular research is more important for what it teaches us about the processes of development than because it may be used for human reproduction.

Every announcement of success with stem cells leads to renewed debates about what to fund, how much to fund, and how to regulate stem cell research. These debates were lively during the 2012 presidential election. President Barack Obama favored stem cell research, including human embryonic stem cell research and federal funding for that research. The Republican candidate Mitt Romney said over the course of the campaign that he favored using stem cells to develop cures but did not support research on embryos. Or rather, he did not support using cloning to create embryos. His position shifted, as did his views about abortion.

These cases—which offer just a few illustrations from one particular sample year, in one particular country, among the many episodes featuring embryos in society—show the divergent points of view about the nature of embryos. These views reflect our divergent ideas about the meaning of life. Science should help to inform policy and public understanding, or at the very least policy should be consistent with and not in conflict with the available established scientific knowledge. Yet the more that embryos come under the microscope, the more persistent the metaphysically based assumptions become about what embryos are, and the greater the conflict grows with scientific knowledge.

Some advocates hang on to metaphysical ideas that they find familiar. Yet it should not be acceptable in modern society simply to cover one's ears and ignore new sources of knowledge; nor should it be acceptable to know nothing about a topic of public concern and to plead ignorance even while holding decided, passionate views about what the policies should be. When the consequences are real, we should embrace learning and refuse to remain ignorant. We should use science to understand the subject, and then use our judgment to make wise decisions about how best to use the knowledge we have. I endorse a preference for reason and evidence over willful ignorance, and I contend that accepting enlightenment is our social responsibility. Policies and laws are a matter of social convention, but they apply to everybody and they have consequences. Therefore, social actions should at least be consistent with the science and should not be driven by hypothetical and metaphysical interpretation of embryos held by a vocal minority when those views are inconsistent with the scientific facts. To see what is at issue, we need to understand the science and how we got to our current knowledge.

Current Policy Context in the United States

This book looks at ideas generally but points to policy in the United States in particular because embryos have taken on such contested political meanings in recent years. A number of other countries, especially those in the European Union, have more nuanced, reflective public policies in this area. By contrast, the United States has a mixed assortment of policies about embryos, which makes it a worthwhile focus for the social discussions in this book. Within the United States, a number of states have passed legislation that regulates or outlaws particular treatments of embryos. Some restrict or specially fund and enable embryo research, while others attempt to limit or to extend abortion rights which are thought to influence our treatment of embryos and fetuses. Not surprisingly, the actions of different states that encourage and even publicly fund research and reproductive choice fall on what is considered the liberal end of the political spectrum. The states that restrict some types of research or seek legally to protect fetuses from abortion at earlier and earlier stages fall on the more conservative end. The National Conference of State Legislatures does an excellent job of summarizing existing legislation at the state level.[4]

Even as the states have been busy enacting legislation, since 1996 federal law still has had only one very general provision in federal appropriations bills to govern the treatment of embryos. Those federal restrictions began in 1996 when two members of Congress, Representative Jay W. Dickey (R-Arkansas, 4th Congressional District, 1993–2000) and Senator Roger F. Wicker (R-Mississippi, 1994–present) opposed federal funding for any type of embryo research. They added a clause (called a rider, because it "rides" along with the original bill) to what is called the Omnibus Funding Bill, which authorizes federal expenditures across many different programs and agencies. The bill provided funding for the National Institutes of Health (NIH), the federal agency that funds medically related research (including research on early development), so the rider applied to all research funded through that bill; Congress has included this Dickey-Wicker Amendment in every funding bill since. The legislation prohibits the use of federal funds to support certain types of embryo research. In particular, "none of the funds appropriated shall be used to support any activity involving: 1) the creation of a human embryo or embryos for research purposes; or 2) research in which a human embryo or embryos are destroyed, discarded, or knowingly subjected to risk of injury or death greater than that allowed for research on fetuses in utero." The legislation refers to title 45 in the *Code of Federal Regulations*, part 46 (45 CFR 46), which covers the protection of human subjects in research generally; with this rider, the congressmen made it clear that embryos must be protected as well. Importantly, the restrictive language identifies a human embryo as if it were well defined, stating that "the term 'human embryo or embryos' includes any organism, not protected as a human subject under 45 CFR 46 as of the date of enactment of the governing appropriations act, that is derived by fertilization, parthenogenesis, cloning, or any other means from one or more human gametes or human diploid cells."[5]

The conditions specified in the legislation, especially artificial parthenogenesis (or the initiation of cell division without fertilization) and cloning (through what is called nuclear transplantation by transferring the nucleus from one organism to another), do not happen under normal conditions, but they can occur in a laboratory or fertility clinic. Biologists consider the results of many kinds of research that combine or eliminate cells to be about embryos. In the federal definition, what is meant by an embryo begins with an "organism." But what, precisely, is an organism here—and in particular what is the beginning point of the organism, namely the embryo?

The *Oxford English Dictionary* (OED) always gives us a useful starting point for looking up the background of words. Its first definition for an embryo states that "the offspring of an animal before its birth (or its emergence from the egg): a. of man. In mod. technical language restricted to 'the fœtus in utero before the fourth month of pregnancy' (New Sydenham Soc. Lexicon) (now, before the third month)."[6] This was also the definition put forth in the first edition of the *Encyclopedia Britannica* in 1771, and it appears in other standard reference sources ever since. An embryo is the earliest stages of a developing organism, and in humans it is the stage up through the first two months or eight weeks of development. At that point, it is called a fetus, because the form is present and all the essential body parts have emerged in rudimentary form.

The definition sounds fairly precise; however, as we have learned from the history of the complex process called development, it is not biologically precise. In fact, in interpreting what seems like clear, basic language, individuals have derived divergent conclusions. Most people believe that the fact that the embryo is biologically continuous with the adult means that all the developmental stages along the way are just different instances of the same individual. But others point out that, biologically, the very earliest stages before the human embryo is implanted in the mother's uterus could be properly called a "conceptus" rather than an embryo. They argue that the full complement of cells will make up not only the embryo but also the placenta that surrounds it, and therefore this early stage is an embryo plus the specialized cells of the placenta. Some call this a preimplantation embryo or a preembryo to distinguish it from the later stages when the embryo has separated from its surrounding placenta. Technically, then, we have a conceptus or preimplantation embryo, then an embryo, and then a fetus; each stage gives rise to the next under normal conditions. The rest of the discussion here will continue to use the common parlance in which an "embryo" refers to the earliest stages up to the point when it becomes a fetus.

This view of a continuum of development stages of an organism makes sense at a material level, but there is a range of possible interpretations about what the existence of the different stages really means beyond biology. At one end, some go on to argue that the embryo is essentially (as in, in its essence) the same as the adult person, so it retains a special continuing individuality or even "personhood" at all developmental stages. Others hold the more moderate view that the embryo may have some special status, but it is not really a person at first and perhaps not for a while—perhaps not even until the eighth week, the point after which the human embryo has rudimentary versions of all the organ systems in place and is then called a fetus. Still others suggest that we should consider the developing organism as not really a full person but as more than "just" matter. At the other end, biologists typically believe that the early developmental stages are just raw matter and do not deserve any special consideration or hold any significant special status at all. Clearly, the material facts give us a solid grounding, but they do not tell us everything about how to interpret what an embryo is in the larger social sense.

Clearly also, much is at stake with these competing interpretations. Now that researchers can perform rich, productive research with embryos, it matters a great deal what we think embryos are and how we think they should be treated. Fertility treatments with engineered and transplanted embryos are a high-stakes business in which it matters a great deal what we think embryos are and whether and how we want to regulate their use. In the United Kingdom, as in some other countries, leaders have conducted wide public discussion and have produced informed regulation for embryo treatment and research through the Human Fertilisation and Embryology Act. As noted, the United States has never had this conversation as a society—it merely has a mixture of widely diverse state legislation, regulations, and court rulings.

Historical Emergence of Divergent Meanings

This book tackles the diverging understandings of embryos and their implications as they play out over time. It traces the development of ideas, focusing on the ways that putting embryos under the microscope has changed our scientific understanding. A central assumption is that getting the biology right should matter for social understanding and for policy decisions.

Although the story focuses on the scientific understanding of embryos, this work must be placed alongside social considerations at major points of inflection. This volume does not give equal time to the social understanding of embryos because that has been widely discussed elsewhere, and there is such a rich literature on bioethics, cloning, stem cells implications, among other topics. Furthermore, this small book is not intended to be comprehensive, but rather to provide focused examples to show the development of ideas and experimental achievements and prospects, as well as to lay out our understanding of embryonic development. As mentioned earlier, my conclusion is that the urgent biological and social implications of embryo research make it all the more necessary that our social understanding of embryos should at the very least not be inconsistent with our biological understanding. This is an epistemological rather than a moral claim. My point is about understanding, and what it means to understand embryos in certain ways. I leave it to others, such as biologist and historian Scott Gilbert and bioethicist and philosopher Jason Robert who are collaborating on just such a project now, to interpret the range of bioethical issues that arise.[7]

The story here proceeds historically overall, but with thematic focus so that the resulting narrative is not strictly chronological. The historical approach shows that some views held today, including those held by former Representative Akin or those proposed in the Sanctity of Life Act, are essentially the same as the typical ideas from centuries earlier. The earliest historical views were hypothetical and necessarily metaphysical, and only over time did the biological understanding develop as the embryos of various species became visible.

The history of biological study of embryos covers two and a half millennia, obviously a rather long time to consider fully in one small book. Fortunately, the ways researchers have approached development can be grouped into several clusters, with various punctuated periods of interest. These periods overlap but can still be characterized as distinct in that they have different emphases and approaches to the embryo.

The historical developments range from a time when embryos stood in for the poorly understood earliest stages of the development of an individual organism, the stages before everything has formed. An animal (or even an idea) can exist in an embryonic form, and it may remain embryonic and never mature to become an adult. For humans, the embryo was largely imagined rather than fully visualized, as nobody had seen living human embryos. As mentioned earlier, only a few animal embryos were accessible to the naked eye. The embryo remained largely a social entity, something that pregnant women may have imagined inside themselves, that the Catholic church interpreted as becoming ensouled (at forty days from conception; after 1869, at conception), and about which most people had only vague impressions. The embryo held a social place standing in for the invisible earliest stages of an organism's life.

Scholars have performed valuable research on the range of ethical and legal understandings of embryos that have resulted from this perspective.[8] As a result, we know that the Torah provides clues to what embryos were thought to be from very ancient times. For the first forty days, the embryo was thought to be unformed and fluid, certainly not yet a "person" in any legal sense. The embryo was, in effect, "like water" at its earliest stages. The Torah also indicates that if somebody caused a woman to miscarry during the first forty days, the only charge would be to pay monetary damages. After the forty-day period, causing a miscarriage was considered murder and was punishable by death. The complexities of such secular and religious views were implemented through law, which reveals that until forty days and the presumed period of ensoulment, the embryo was not seen as having its own moral or legal status.

Early Catholic authorities held similar views, though current Catholic leaders are not as keen to recall this because the church's views have changed. In the fourth century, St. Augustine expressed his opposition to abortion, which he regarded as breaking the important moral connection between sex and reproduction. Yet he also seemed to regard the early stages of development as unformed and as therefore having a different status. St. Augustine is frequently quoted as having said that the early developmental stages often perish like seeds before they have been "fructified."[9] Abortion was a sin, but it was not homicide in this worldview.

In the thirteenth century, St. Thomas Aquinas reasoned similarly.[10] For Aquinas, the early stages of development allow growth, but there is no ensoulment and no form as yet. Following Aristotle's thinking, he saw the early stages as involving a vegetative "soul" and then the organism develops an animal soul, which brings the beginning of form. Only with the presence of all the body parts and the appearance of the human rational soul did the form count as having become fully "hominized." Only then would its destruction be considered homicide, because only when the three souls—vegetative, animal, and rational—were working together would it be truly human. Pope Gregory XIV in 1591 explicitly declared that ensoulment occurs only later and not at the beginning of development.

This particular metaphysical interpretation of "delayed hominization" persisted until 1869, when Pope Pius IX was involved with the conservative reforms of what was called Vatican I.[11] Pius IX proclaimed that life begins at conception, which he interpreted to mean that an individual life begins at the moment of fertilization. The pope explicitly declared abortion as an offense worthy of excommunication, which brought clarity and tightening of the Catholic restrictions on reproduction and all forms of controlling birth.

This action by Pius IX immediately raises the question of why he was so committed to this point of view about early human developmental stages. Had he perhaps learned of the recent biological research that showed the nature of fertilization of cells? Or was he, as commentators have suggested, particularly eager to assert the church's control over its members and to expand the number and nature of prohibited behaviors? We can only conjecture, but scholars have provided no evidence that he was driven by knowledge about what was going on in biology, nor that he was addressing the matter from the point of view of reason. Rather, as is perfectly appropriate, faith, church doctrine, and expediency seem to have driven Pius IX.

One additional example of understanding embryos that is not quite metaphysical and not quite biological is the mother's point of view. Women can experience but not directly see the embryos inside them, so there is often a unique kind of conviction about what is there. Their views may be grounded in faith and feelings, and may be more or less informed by an understanding of what is happening biologically. For most of history, when women became pregnant, they may have realized it during the embryonic stages of development but more often they did not. Later, they might sense some version of "quickening," defined as the point when a woman first feels the growing embryo/fetus inside her—quickening referred to the moment when the organism became alive. Women may have ascribed this moment to "ensoulment" if they held to particular religious views. The mix of experience with biological and social cues has created a special understanding for mothers.

Consideration of a mother's view does not play a further role here, except to remind readers of the growing interest in (and pressure on) mothers taking care of the embryos growing inside them. A rich literature explores the various convictions over time that developing embryos are influenced by the mother's environment, by what the mother eats or drinks or smokes, by whether the mother has certain kinds of thoughts or listens to certain kinds of music or performs certain kinds of activities. At various times, it has been thought that the sex of the child was determined by which side the mother slept on, or the embryo's health depended on whether the mother had healthy thoughts. Most recently, what the mother eats and drinks, along with her environmental conditions are thought to influence embryonic development and gene expression, which might seem to place further pressures on the mother. Throughout history, this implicit responsibility has reflected and generated interactions among feelings, belief, science, and medicine, which bears much further study than it has received.[12]

Structure of This Book

Any careful look at embryos should start with the first, hypothetical period and categorize the history of studying embryos into seven somewhat overlapping, roughly chronological periods. These include the *hypothetical embryo,* in which theory and philosophical and social factors played an important role; the *observed embryo,* which brought an intense interest in observing developing forms under magnification, accompanied by debates about what people were seeing; the *experimental embryo,* in which experimentation was added to observation to gain more additional knowledge about how embryos work; the *inherited embryo,* which focused on the nucleus, DNA, and inheritance; the *evolved and computed embryo,* in which computational models brought new understanding of the processes and constraints of development; the *visible human embryo,* in which human embryos in Petri dishes could be studied directly; and the *engineered/constructed embryo,* in which researchers engineer and build according to their own guidelines and are no longer restricted by "normal" development as before.

Chapter 1 lays out the questions for the book and for those studying embryos, pointing to the different meanings that have been and are now being assigned to embryos, and also to the contexts in which those different meanings have arisen. This chapter sets up the recurring questions for the volume as well as the approach. The book proceeds through different historical, though not strictly chronological, episodes before returning to consider the implications in the concluding chapter.

Chapter 2 introduces the first hypothetic period of biology, starting with Aristotle and carrying us to the end of the nineteenth century with the observed embryo. As with so much of the history of science, Aristotle was the first recorded observer of the generation of animals. He gave us the hypothetical embryo and began to ground his understanding in empirical observations as well, though he had no microscopes to help him. Aristotle observed chicks developing in the egg, and he offered explanations of what was going on. In his work, *Generation of Animals,* he provided us with ways of thinking biologically about development. Others then built on Aristotle's ideas, observing what they could in animal species. Watching, comparing, interpreting, and experimenting, researchers have sought to understand what is happening inside the developing organism, asking questions such as what forces or factors drive development? Taking Aristotle as an example tells us much about how scientists develop their scientific understanding within changing and often challenging social contexts, and examining biological research makes up the major part of the following chapters.

Chapter 3 introduces experimentation, as researchers began to manipulate eggs to see more and to answer questions about the patterns and processes of development. This period takes us to the 1950s, a time when women understood that the embryonic form needed to be nurtured. Dr. Benjamin Spock and the general public health movement called for taking care of the embryo. Only a generation earlier, in the 1920s, women had learned minimally that first there is sex, then one gets sick, then after some months a baby is born. They did recognize that along the way something called an embryo came into being, but the idea of an embryo had no biological or tangible meaning for them other than as something not really fully "there" yet.

Chapter 4 brings in genetics, with the ideas of inheritance and evolution as well as early ideas of computational approaches to understanding genetics and development. This period had its foundations earlier in the twentieth century, and it continues today with important work at the intersections of developmental biology, evolution, and computation. The period of the 1950s through the 1980s introduced many new ideas and approaches.

The 1970s brought a mix of biological and social developments that are laid out in Chapter 5, which introduces the visible human embryo. This period brought with it the women's movement, which highlighted what happens in women's bodies, as well as an increased awareness of abortion, which led to *Roe v. Wade*. The technique of *in vitro fertilization* (IVF) also first appeared in this period. The socially understood human embryo began to be a biologically studied embryo, like the embryos of other species: a biological object begins with fertilization and continues through a predictable process of development, growth, and differentiation to the point when the embryo takes on the form of a human being, with all the organ parts in place and beginning to function. This latter stage occurs at eight weeks (or just a little more than those forty days considered essential by the early church fathers and other religious groups and philosophers).

In 1978, with IVF, the earliest stages of the human embryo were no longer left to the imagination—they took place right in a glass dish for anybody to see. The embryo clearly starts as a single cell, which is then fertilized by a sperm cell, begins cell division, and undergoes continued divisions to produce more and more cells. It must be implanted into a woman to receive nutrition and deposit wastes, so this embryonic developing biological thing is no longer independent and autonomous. Biological data from other species as well as novel imaging technologies began to create a picture of what happens at each developmental stage.

The socially imagined embryo had, by the middle of the twentieth century, been joined by the biological object: the physically imaged embryo. The social embryo persists, however, especially vividly through abortion debates, as illustrated by well-orchestrated demonstrations set up on university malls with abortion protestors in t-shirts declaring that they are "former embryos" along with the message that abortion of fetuses is killing "babies." This socially imagined embryo is seen as a miniature human being, with its own individuality and integrity and just waiting to grow up and become a person. The biological embryo, by contrast, is one that develops gradually, in which the earliest stages are understood to be less fully formed and less developed than later stages. These two embryos coexist in the public realm, sometimes occupying the same minds while playing different roles.

Chapter 6 presents the early ideas of engineering and constructing life, initiated most energetically by Jacques Loeb in the late nineteenth and early twentieth centuries. Loeb offered an ideal and a set of goals grounded in his assumption that life is mechanistic and can, in fact, be engineered. His own effort with artificial parthenogenesis and physicochemical interpretations of fertilization inspired others, but his ideals ultimately ran ahead of reality. It took decades more before science and engineering advancements caught up with Loeb's imagination.

Chapter 7 discusses ways that researchers have shown the extent to which it is possible to go beyond the ideal and the hope in constructing embryos. Although most have heard something about stem cell research, most of the public does not yet fully understand the range of just what is possible already or why it might matter. But biologists have shown with various species their capabilities for manipulating embryos. For example, they can hollow out cells, then reengineer them with new DNA, new mitochondria, cell structures, or recombined nuclear material. Then as the cells divide, it is possible to take some cells away or add new ones (up to at least the eight-cell stage). The result, a different embryo than what would have developed, is highly engineered and even constructed by the researcher.

That chapter then looks at stem cell research, sorting out the differences between embryonic and adult research, and introducing induced pluripotent stem cell research (iPSC) and the debates about the efficacies and limitations of each. Stem cell research has taken on a life of its own for social reasons, but biologically understanding stem cells and the way they move from being undifferentiated to becoming different kinds of cells gets at the most fundamental issues of developmental biology.

Chapter 8 looks forward. Most of the research described so far has been done with nonhuman species, and one might well ask whether anybody is actually carrying out the engineering work on humans. Why would they want to? The answer lies with the desperate hope for clinical therapies. If we could cure terrible diseases by engineering the embryo, many would argue that we should try; others would nonetheless maintain that we absolutely should never do so if it involves interfering with embryos. Of course, there are many questions about why we think we can or should address diseases in the embryonic stage, but we will come back to those later. The point here is that a tremendous amount of engineering is possible, and some is probably desirable under certain conditions. A great deal is already happening in various species, so we, as a society and individually, should understand what is happening and draw thoughtful conclusions.

The look to the future includes reflection on attempts to create new life through synthetic biology. This research occurs in the context of thinking about complex adaptive systems through systems biology. The work requires understanding of the way that heredity and development work together in the environment through what is called epigenetics.

Rather than trying to provide complete coverage of every possible contribution or every possible idea (which is impossible anyway), this narrative selects examples to demonstrate what is at issue and what is important in each period. Therefore, some examples appear in depth, and others receive only brief introductions. The final chapter, entitled "Therefore," offers interpretations of the implications of the story told here and reflections to carry us forward wisely and equipped with knowledge and understanding into a changing world of understanding embryos under the microscope.

2

Hypothetical and Observed Embryos with Microscopes at Work

We go back here to that initial time when people could not actually see much and had to imagine what happened in the earliest stages of development. This chapter provides a brief description of the first two historical periods of embryo studies: the hypothetical and observed embryos, examining what those periods entailed. One interpretive theme runs throughout, offering two ways of imagining what was going on with embryos: epigenesis versus preformationism. In general, these two ways of imagining development remained distinct and non-overlapping. Indeed, during most of history the two were in direct competition. Yet it is possible to hold a view that brings together aspects of both positions, and developmental biology today seeks to do just that. It is worth learning more about these two as distinct approaches as well as the ways and reasons why they have come together in more recent decades.

Epigenesis arose as a clearly articulated view about development, wherein embryonic form emerges only gradually. In fact, an individual begins as a mixture of fluids, according to many epigenetic interpretations such as Aristotle's. Male and female fluids come together through sexual intercourse, and they begin to combine and work together to generate form. These epigenetic interpretations held that the early developmental stages are thus entirely unformed at first, yet they have the potential to become fully formed through some process about which considerable disagreement persisted.

Epigenesis allows a materialistic emphasis on the physical stuff of development, but it requires something to drive the process from unformed material to a complete and functioning organism. Before the late nineteenth century, those thinkers who wanted to have some causal explanation for development rather than relying on mystical unspecified changes had to appeal to an end-directed teleological cause. Such a cause could, in effect, pull the organism from its amorphous inception toward its final goal of becoming a fully formed adult of the right kind. This reasoning started from the epistemological conviction that having a material, nonmystical causal explanation is important for understanding what embryos are. Others who held the same views hypothesized the existence of a special kind of vital force that could serve the same purpose of driving the organism in the right direction.

Aristotle accepted some version of both ideas with his notion of "entelechy" to capture the idea of an object that is fully actualized as a result of the causal process. By the eighteenth century, his idea of entelechies gave way to alternative interpretations until it was picked up again around 1900 and modified for modern purposes by the epigenesist philosopher Hans Driesch. Eventually, however, epigenesists set aside teleology and entelechy and embraced the idea that the material itself could drive the process, and that the process required neither a pull nor a push because of special forces. In any case, according to the epigenetic view, the resulting development of the individual is driven internally but is also a product of its interaction with its surrounding environment.

In contrast, the preformationist view that appeared in the seventeenth and eighteenth centuries initially suggested that the organism's form was there from the very moment of inception, just hiding or so small that the human observer could not see it yet. This literal version of preformationism gave way to various forms of predeterminism by the end of the nineteenth century, according to which the formed structure was not literally already there from the start, but nonetheless already determined or predestined in some sense. Those who held this view saw the organism as having some information or organization available from the beginning that caused the organism to follow a path already laid out in some way.

Only by the late twentieth century did developmental biologists begin to work out a more sophisticated understanding of how some type of inherited predeterminism through genes could interact with responses to the surrounding conditions that form the environment for each embryo. They began to provide explanations of how resulting interactions could allow form to emerge gradually and through a process that starts with something that was not yet a fully formed organism in any sense. In modern terms, heredity through chromosomes and genes has brought predetermination of a sort, except that according to the modern interpretation, the expression of the genes responds to environmental conditions both internal and external to the embryo itself. This new view of development as a dynamic interaction between genes and environment accepts both epigenetic development and a predeterministic starting point that decides some features but only influences others.

Until the recent compromise that embraces both views in modified versions, however, debates raged between epigenesists and preformationists. These debates often involved other central underlying assumptions about how life works, drawing on theories of vitalism and materialism. Those thinking about development started with hypotheses about human embryos, which they could not actually observe, then adjusted those ideas to account for observations of other types of embryos of diverse sorts. In all cases, they interpreted embryos as having different meanings. Today, genetic determinists still argue with developmental biologists about which factors are the most important in shaping the details of a developing organism. This chapter looks in more detail at the history of thinking about these alternative interpretations and the efforts to observe developing organisms to gain a better understanding of embryos.

The Hypothetical Embryo

Aristotle is worth spending some time understanding because his ideas prevailed for nearly a millennium and a half. Aristotle lived from 384 to 322 BC, a time of lively Greek thought. A student of Plato's, Aristotle was the son of a physician who served King Amyntas in Macedonia. Aristotle himself served King Philip and then became the teacher of Philip's son Alexander the Great. Aristotle grew up within the aristocracy, and he had the luxury of time and leisure to observe the world and think about it. Many people had a similar luxury and never took the opportunity, so Aristotle is distinguished by his drive to explain everything—really, *everything*. He explained how the heavens move in a regular and predictable way, as they surely do, as well as what composes the structure and nature of the universe and how the earth is different from the heavens. Embryos and life itself were but one part of that complex Aristotelian universe.

Aristotle saw us humans on earth, surrounded by animals and plants, as part of the sublunar world, the changing world that exists between the moon and the earth. For him, we live at the center of that universe, and that means that all changes that happen beyond us as well as the conditions immediately around us can affect us. The heavens beyond the lunar sphere, which he imagined as a sphere around the earth that included the moon, he regarded as perfect and predictable. Stars and planets and the sun go around and around in a heavenly way that is also perfect and predictable. According to Aristotle, the heavens consist of ether, which is a perfect substance and is embedded in spheres that carry the motion for the stars and planets.

But Aristotle also concluded that the motion of all those heavenly spheres turning around the earth necessarily stirs things up here at the center. And all "things" on earth are made up of elements and what he called their qualities, or what we might call the characteristics of hot, cold, wet, and dry in various combinations. The four basic elements of our earthly central world include earth, fire, air, and water. These are connected to the qualities so that a particular material object might be a mix of earth and water, say, that is hot and dry—or any of the other possible combinations. The motion of the spheres that carry the moon and the stars out beyond the moon causes a mixing up of those elements in the area below the moon. This in turn causes change, which Aristotle thought of as "generation and corruption." The physical world changes, and so do the animals and plants. For each individual organism, the changes involve a process of moving from what is just potential but as yet unrealized or undeveloped to a state of being actually developed. In this view, "generation" involves the actualizing of the potential, which yields life and then eventually senescence and death as part of the natural processes.

In addition to the four elements and four substances, Aristotle argued that four causes guide all change.[1] He offered his own general account of the four causes, which remains general in the sense that it applies to everything that requires an explanation, including artistic production and human action. Aristotle recognized four types of causal answers to a "why" question, where all four of the causes work together.[2] Why does a particular result occur? Because of (1) the material cause: "that out of which," or the material of which the thing in question is made; (2) the formal cause: the form or shape of the thing in question; (3) the efficient cause: the primary source of the change or lack of change, which includes the actor who brings the change, as well as the process; and (4) the final cause:

"that for the sake of which a thing exists" or the goal for the thing in question.

What does this mean for living organisms, we might well ask, and particularly what does it mean for humans and for their development? Scholars have offered various detailed interpretations, but for human development the basic idea is that material causes begin with the "menstrual fluid," final causes with the actualized adult form of the human, efficient causes with the "seed" or sperm, and formal causes with the essence or form.

Aristotle himself suggested that the formal and final causes might be the same thing for man, with the goal of an individual's life being to realize the actual adult living form. In fact, the only major difference between the man who is living right now and one who has just died a minute or so ago is not that they look different, because they typically do not, but rather in the case of the dead man the final cause has ceased to act. "That for the sake of which the thing exists" has failed to exert any causal force.[3]

Aristotle's boldness and imagination in explaining the universe and all its parts are quite astounding, and they are even more impressive because from our modern perspective many of his views were not entirely wrong. That is, Aristotle's interpretations for how things work make great sense, even when the details of his particular explanation are replaced by other, more complicated ideas from our more extensive knowledge about the world. We no longer regard the heavenly bodies as traveling around on celestial spheres, nor do we place the earth at the physical center of the universe, and our concept of elements and causes has become far more complex. Yet there is something about the consistency and regularity of Aristotle's views that still makes sense and still largely works.

For embryos, Aristotle started in the right place. He obviously realized that women go through monthly cycles and lose menstrual blood on a regular and predictable basis. He also presumably realized that sexual reproduction involves a mingling of fluids. Therefore, it made perfect sense to build on the empirically observed and interpret the mingling of fluids from the male and female as a starting point for development. Given the cultural assumptions of his time that males are more important than females, it also made sense to assume that the female provided the raw material for development and the male provided something more important, the "seed" or efficient cause to carry out the process of development from raw material into the right kind of product.

Because human mothers always give birth to other humans rather than some other kind of animal, and other animals always seem to breed true to their types as well, it made good sense also to assume that there is a formal cause to direct the particular pattern and form. And the goal of each life was to actualize the potential of being a human person—that is, to achieve and express its essence. That essence is made up of a balance of elements and qualities. Though the list of four elements, four qualities, and four causes might seem too simple, it all fits together so nicely and provides an excellent starting point.

Aristotle had more to go on than just observations of human reproduction and theory, however. He was always a keen observer, and in the case of embryology he watched chicks developing. Chicks seemed to him clearly to start as unformed masses of material, and only gradually to develop the form that defines them at the point when they hatch. In fact, Aristotle managed to observe the moment when the heart begins to beat. The gradual emergence of form out of the unformed seemed so obviously correct that epigenetic interpretations of development prevailed even after the Scientific Revolution of the seventeenth century presented alternatives.

It is a wonderful thing to watch the form emerge before your eyes, as Aristotle did.[4] First there is what appears to be simple material; then it forms into an organized red shape; then it beats, and the blood begins to flow. Aristotle presumably saw the same thing and sought to explain this remarkable phenomenon in the material scientific terms of his comprehensive natural philosophy.

Aristotle's interpretation of development was thus grounded in both observation and his theoretical view of the universe. His embryo was necessarily partly hypothetical. Yet he also drew on his experience and observations, which led him to an empirically supported view of epigenetic gradual development. At that time, the view of the biological embryo corresponded closely to the metaphysical views, of which all the leading versions held that the life begins at forty days. That is, everyone agreed with the gradual epigenetic emergence of form over time. The difference was that the theologians generally did not assign a material or scientific explanation to the phenomenon, but Aristotle did.

The Observed Embryo

Aristotle did observe some animal development, such as the chick in the egg, but he had no microscope to help him. By the seventeenth and eighteenth centuries, the Scientific Revolution and the Enlightenment, many more people had begun to observe nature closely, and they began to use microscopes to help them see better and more. The idea that one should look to see for oneself became a theme for the new era of science. Rather than taking the word of authorities, whether Aristotle or the Catholic church, one was encouraged to examine the evidence and interpret the phenomena of nature with new eyes—specifically, with one's own eyes.

William Harvey took up the challenge. A British surgeon who had studied at the University of Padua in Italy, Harvey most famously discovered that blood circulates rather than being constantly being used up by the body (as his predecessors such as the second century Greek physician Galen had famously thought).[5] Harvey confirmed the circulation theory with careful observations, experiments, measurements, and a highly sophisticated scientific approach to an old problem.

Harvey also asked how generation occurs, and here too he looked carefully and then developed interpretive theories that he tested with further observations. In particular, Harvey looked deep inside a female deer, just after copulation. He did not directly see any eggs or semen, nor did he find the mixed fluids that Aristotle had mentioned. Even so, he remained an Aristotelian epigenesist and empiricist—he believed that what we have called the hypothetical embryo was not just hypothetical but the reality. Harvey went on to hypothesize that every animal begins as an egg ("Omne vivum ex ovo"), and he believed that these eggs resulted from a process of conception. He did not know the details, but observing chick eggs persuaded him further that the process of formation of a new organism is epigenetic and gradual, beginning from a defined starting point in the hypothetical egg.

In light of the historical evidence, we can conclude that Harvey did not actually see what we call the "egg" in the deer, but he did probably observe the amniotic sac or other structures connected with reproduction. Harvey thought that his egg underwent development thanks to a "formative virtue," which was consistent with Aristotelian causes. This virtue added a causative influence that served as a special vitalistic life force applied to the material starting point. Although his egg remained just hypothetical, his later observations confirmed his emphasis on a localized, defined, material starting point for individual development.

Peering at early developmental stages with the naked eye can show some details in some species, but microscopes reveal far more details in even more species. Harvey used a simple lens for his observations, as did other observers; in the Scientific Revolution, microscopy using compound microscopes would quickly become part of embryology. Microscopes seemed to offer a way to "see inside" something very small, and perhaps offered a valuable tool for resolving philosophical debates.

By the eighteenth century, new observations and new theories were sparking heated debates. The assumption that development occurs epigenetically and that form only emerges gradually remained the dominant view.[6] With only a few exceptions, such as the materialistic mechanist René Descartes, who emphasized the role of matter in mechanical motion, epigenesists adopted some version of the vitalistic interpretation and thereby assigned a distinct meaning to life. They reasoned that if the whole structured form was not there at the very beginning, then this raised serious questions about how the form could possibly emerge out of the unformed. There simply had to be some driving vital force or causal factor, they assumed. Otherwise, how could the process work so reliably to produce adults of just the right sort in case after case?

Confirmed epigenesists such as Caspar Friedrich Wolff from Berlin had no problem with such a gradualistic interpretation. As Wolff reported in 1759, he had looked in particular detail at chick eggs and saw what Aristotle had seen: unstructured material that gradually takes on form and becomes structured. Wolff was convinced that what he had observed was an accurate demonstration of what exists: an epigenetically emerging heart.[7]

But other people had different ideas. For example, a contemporary of Wolff's, the Swiss naturalist Charles Bonnet, also looked upon the developing chick and its developing heart, and arrived at a different interpretation. In his work published in 1769, Bonnet noted that he could not see the fully formed heart at the early stages either. But instead of assuming that because he could not see it, it was not there, Bonnet concluded that it simply had to be there. He offered that microscopes might not be strong enough yet to see it, and he was willing to await better technology. Or perhaps the heart remained too small or too subtle to be seen, but that would be a limitation in our powers and not a proof that the structure did not exist.

Bonnet adopted this suggestion because he was first of all a materialist, who believed that he must explain everything in terms of matter and motion, and he rejected any sort of unseen vitalism. He rejected the idea that form can arise from unformed matter because that interpretation seemed to demand some sort of vital force or principle to guide formation. And vitalism just did not make sense to Bonnet, who started with a rational and materialistic view of the world. Given his rejection of vitalism for metaphysical reasons, Bonnet adopted a preformationist view and rejected epigenesis. During the eighteenth century, various versions of preformationism coexisted and competed with what had been the dominant and almost uniform commitment to epigenesis.

Discussions about the embryo therefore got caught in the crossfire about underlying assumptions for these eighteenth century figures. Amid their observational work, they interpreted what they could see—and also what they could not see—in different ways. Some considered that what they could see remained insufficient to provide either a clear description or an explanatory understanding of the phenomenon in question. In the eighteenth century, when confronted with uncertainty, the research community kept looking at more and more organisms to interpret what happens in development. Epistemological convictions about the best way to gain knowledge (namely, through direct observation) played a central role in the discussions about embryos, as it did about other aspects of life and nature.

Abraham Trembley was one of these curious observers. This Swiss naturalist was especially fascinated by the hydra, a form of simple freshwater polyp that can regenerate when it is injured. Trembley thought he had discovered a new species (though in fact others had already recognized them), and he published his findings in 1744.[8] The hydra's ability to regenerate is quite remarkable, as is the fact that it does not seem to age or die as part of its natural life. Chop off its head, and it will grow back a new head—or more than one, which is probably why they were named after the ancient mythical beast with multiple heads. Chop off other parts, and those regenerate as well. This power of regeneration raised such fundamental questions about what counts as an individual organism and how it responds to injury that researchers are still exploring this phenomenon and asking the same questions all these centuries later, though now their research encompasses the lively fields of regenerative biology and medicine.

Von Baer, Eggs, and Cells

No matter how careful the observations, simply recording more and more descriptions of unusual stages in a variety of organisms cannot settle all the core questions about how development occurs. Even when researchers saw essentially the same events, they might interpret them differently, as Wolff and Bonnet had. Into this swirl of activity, the Estonian (or Russian) embryologist Karl Ernst von Baer set out to determine step by step what happens from the beginning of development. He looked at chicks, as Wolff and Bonnet had, and determined by watching very closely that the form definitely arises slowly and gradually, with the egg undergoing largely predictable movements over time. He mapped out the changes in each step, and suggested what forces might be at work to make the changes occur in that particular way.

Von Baer also wanted to know what happens at the very beginning in mammals, where the development happens inside the mother and is therefore out of sight. In 1827, he reportedly visited his physiology professor Karl Friedrich Burdach, in Dorpat, Russia, and found the Burdach's family dog in heat, meaning she was biologically ready for mating. Von Baer and Burdach both wanted to know what they would find if they looked inside her. Would they find nothing?

Or would Harvey's hypothetical egg be there as the actual starting point for development?

They sacrificed the dog, opened her up, and looked inside. They did, indeed, find an actual physical and observable, even if very small, egg. This led von Baer to conclude by inference that in every case where an animal begins through sexual reproduction by a male and female, the offspring begins with an egg ("ex ovo omnia").[9] Von Baer had just discovered the mammalian egg, which led him to the idea that other mammals also begin as eggs.

Von Baer then turned to careful observation of the stages of development in frog eggs, which proved especially valuable for studying development. Frogs deposit a large number of very large eggs into a pool of water, which makes it easy for the observer to scoop up a batch and take them home. Then he could watch them develop—first one cell, then two, then four, and so on. Some species have relatively transparent eggs, and most are quite easy to observe. In 1834, Von Baer laid out the developmental steps very clearly in pictures so that any reader could see just what happens at each cell division.[10] He went through the same process with chick development (Figure 2.1), following Aristotle's observational pathway.

Notice that something new has sneaked into the discussion that had not been there before: cells. You might ask, where did those come from? Or you might not ask because you take for granted now that we know that organisms consist of cells, and that the earliest developmental stages involve cell divisions. But before the early nineteenth century, it had not been obvious that life consists of cells. In fact, a good deal of controversy persisted about the matter. When the British microscopist Robert Hooke first observed cells in cork and began to suggest that they are more widespread as well, others resisted this interpretation. It took a convergence of the ideas that there is a basic unit of life and that the eggs at the beginning for each life are those cells, and of significant improvements in microscopy to establish cell theory in 1838. The Germans Matthias Schleiden,

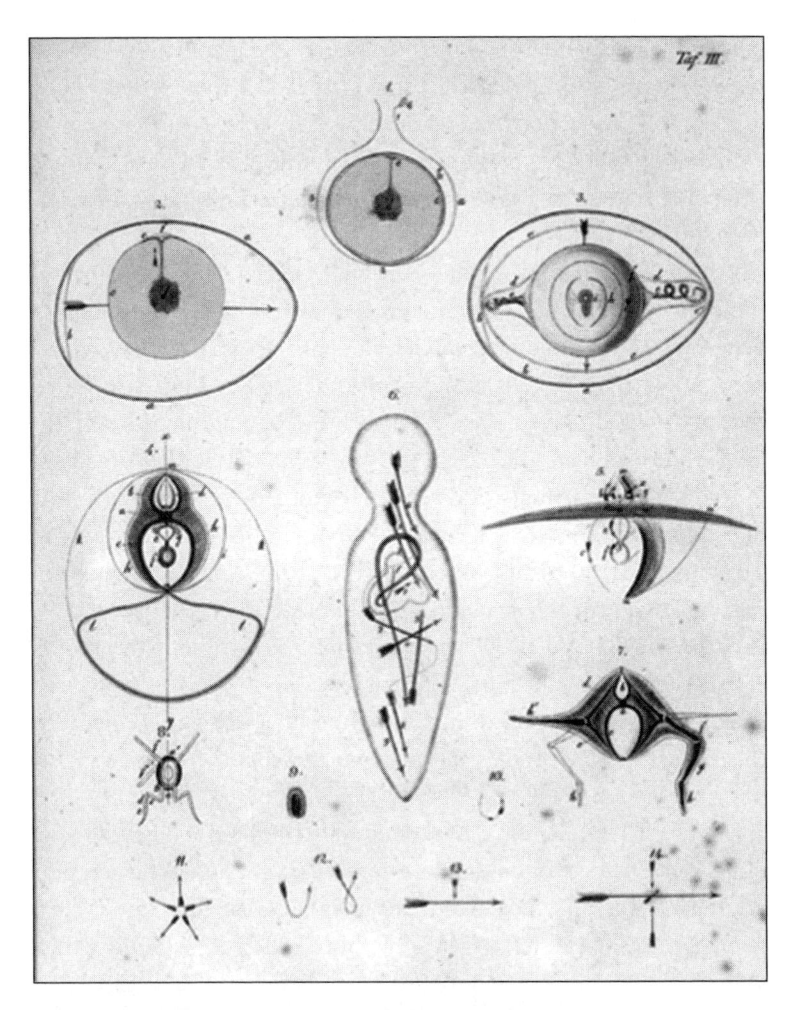

Figure 2.1. Chick development described by Karl Ernst von Baer. Von Baer studied the patterns of morphological change in the early developmental stages. He compared the pattern of chick development with those of other organisms. From von Baer's *Über Entwicklungsgeschichte der Thiere: Beobachtung und Reflexion,* Part I (Königsberg: Gebrüder Bornträger, 1828), Plate 3.

studying plants, and Theodor Schwann, studying animals, provided the foundations. Together they put forth a cell theory that, while not entirely accurate from our perspective, focused discussion in valuable ways on cells and questions about their various roles. Some version of cell theory has served as a grounding for biological science ever since.

Others quickly took up the idea that cells are the fundamental building blocks of life. First Robert Remak and then the even more vehement and well-connected German pathologist Rudolf Virchow proclaimed that all cells come from other cells; others concluded that all living organisms begin as cells. By the mid-nineteenth century, observers like von Baer were seeing the developing organism in terms of cellular units of life. Furthermore, by the late 1840s evidence had accumulated that each individual life begins as an egg cell that is fertilized by a sperm cell to form a new combination that begins a process of cell division that leads to a complex organism. Martin Barry led the way to seeing the various parts of the cells as important; in particular, he saw the nucleus as a defined structure inside the cell, though it would take a few more decades to sort out its role or even its structure through the various stages of cell division.

Within this cellular view, life comes only from life and never spontaneously grows from nonliving matter. Spontaneous generation does not occur, and those who had imagined that they saw it happening must instead have seen something else. If they thought they saw maggots crawling out of meat, as Francesco Redi did, then surely some flies must have settled on the meat earlier and deposited eggs. If they thought they saw worms spontaneously generating out of soil, an egg must have provided the starting point, or else the worm regenerated from another worm that had been cut in half. The improved achromatic microscopes (so-called, because their lenses effectively avoided the rainbow-like chromatic fringes that had plagued earlier microscopes and had distorted observations) helped confirm the theory.

Embryos and Evolution

One additional piece of the story brings in alternative meanings of how individual organisms and individual species fit into the bigger picture of life: evolution. For Charles Darwin, embryos and evolution were closely tied together. Evolution shapes the capacity of organisms to respond and to develop in particular ways. So although most people think of evolution as something very different from heredity and development, and very distant in time and effect, every one of us carries the remnants of evolutionary inheritance and adaptation, and all these processes work together. Darwin shows connections from development to evolution and from evolution to development, as do the modern movements of "evo-devo" or the alternative conceptualization as "developmental evolutionary biology."

In chapter 13 of his 1859 work *On the Origin of Species,* Darwin pointed to embryos as his favorite evidence for evolution by natural selection.[11] He had observed embryos of many species, and he saw that they look most alike in their earliest stages. They then diverge, as the later stages bring differentiation of cells and of types (or species) of organisms. This progressive differentiation suggested to Darwin that the early stages of development actually come from the ancestral form; the embryos and later stages result from adaptations to changing environmental conditions. Therefore, looking at similarities and differences among embryos of different species would help determine the historical evolutionary relationships among those different species. The fact that the early stages remain so similar across so many species provided very strong evidence for Darwin's evolutionary claim that they are descended from the same ancestor through natural selection. Clearly, embryos played a very important role for Darwin's argument.

In his typical Victorian prose in the chapter he called "Recapitulation" and the section on embryology, Darwin asked,

> How, then, can we explain these several facts in embryology,—namely the very general, but not universal difference in structure between the embryo and the adult;—of parts in the same individual embryo, which ultimately become very unlike and serve for diverse purposes, being at this early period of growth alike;—of embryos of different species within the same class, generally, but not universally, resembling each other;—of the structure of the embryo not being closely related to its conditions of existence, except when the embryo becomes at any period of life active and has to provide for itself;—of the embryo apparently having sometimes a higher organization than the mature animal, into which it is developed.[12]

He then concluded that "I believe that all these facts can be explained, as follows, on the view of descent with modification."[13] He also added,

> The leading facts in embryology, which are second in importance to none in natural history, are explained on the principle of slight modifications not appearing, in the many descendants from some one ancient progenitor, at a very early period in the life of each, though perhaps caused at the earliest, and being inherited at a corresponding not early period. Embryology rises greatly in interest, when we thus look at the embryo as a picture, more or less obscured, of the common parent-form of each great class of mammals.[14]

In at least some important ways, Darwin built on his contemporary von Baer's laws of development. For von Baer, the fact that embryos seem similar in the beginning reflected the relationships among the various broad types. Rather than seeing a general ancestor for all life, he saw five different large groups with different creation points at the start. In his volume on the development of different types of organisms in 1828, von Baer laid out a set of *Scholia,* later termed laws for development. It is worth giving these in von Baer's own (translated) words because the list has been retranslated and re-interpreted many times with varying results. Von Baer said:

1. That the more general characters of a large group of animals appear earlier in their embryos than the more special characters.
2. From the most general forms the less general are developed, and so on, until finally the most special arises.
3. Every embryo of a given animal form, instead of passing through the other forms, rather becomes separated from them.
4. Fundamentally, therefore, the embryo of a higher form never resembles any other form, but only its embryo.

> The history of the development of the individual is the history of its increasing individuality in all respects.[15]

Despite the fact that they saw similar patterns when they looked at embryos, up to his death in 1876 von Baer never accepted Darwin's ideas on evolution by natural selection; rather, he envisioned a past point of creation for the different types and divergence from there. For von Baer, different types each have a different origin, and any divergence in developmental details occurs only in the details and not in the fundamental pattern. Darwin, of course, saw development in his own terms of evolution, with divergence because of adaptation after descent from a common ancestor.

The creative German researcher Ernst Haeckel shared Darwin's view. Historical scholarship shows us the romantic Haeckel, in love with life and with his wife. Yet Haeckel was struck by personal and professional tragedies that shaped his own life and career, as historian Robert Richards discusses so compellingly in his book on *The Tragic Sense of Life.*[16] Haeckel observed nature and saw the colorfulness of nature's diversity. He came to see diversity in terms of something like Darwin's "view of life," which Darwin had noted on the last page of the first edition of *On the Origin of Species,* "It is interesting to contemplate an entangled bank, clothed with many plants of many kinds, with birds singing on the bushes, with various insects flitting about, and with worms crawling through the damp earth, and to reflect that these elaborately constructed forms, so different from each other, and dependent on each other in so complex a manner, have all been produced by laws acting around us." In this frequently quoted last paragraph, Darwin continued, laying out what those laws of nature are. He ended with the exultation, "There is grandeur in this view of life, with its several powers, having been originally breathed into a few forms or into one; and that, whilst this planet has gone cycling on according to the fixed law of gravity, from so simple a beginning endless forms most beautiful and most wonderful have been, and are being, evolved."[17]

Haeckel shared this sentiment, and he boldly added his own law of recapitulation, called the biogenetic law, according to which "ontogeny recapitulates phylogeny." That is, development of an individual roughly follows the evolutionary history of the group or species to which that individual belongs. Haeckel gave his readers beautiful and compelling images of different species at early developmental stages, showing that the earliest stages are the most alike (Figure 2.2). A very fine naturalist, Haeckel had an eye for the beauty and patterns within the apparent messiness of nature. One thing Haeckel saw was close similarities among the early developmental stages of embryos of different organisms. At the time, his evolutionary suggestion that embryos closely resemble each other because they share a common ancestor was highly controversial, yet its tidiness appealed to those who accepted the importance of evolution. Ever

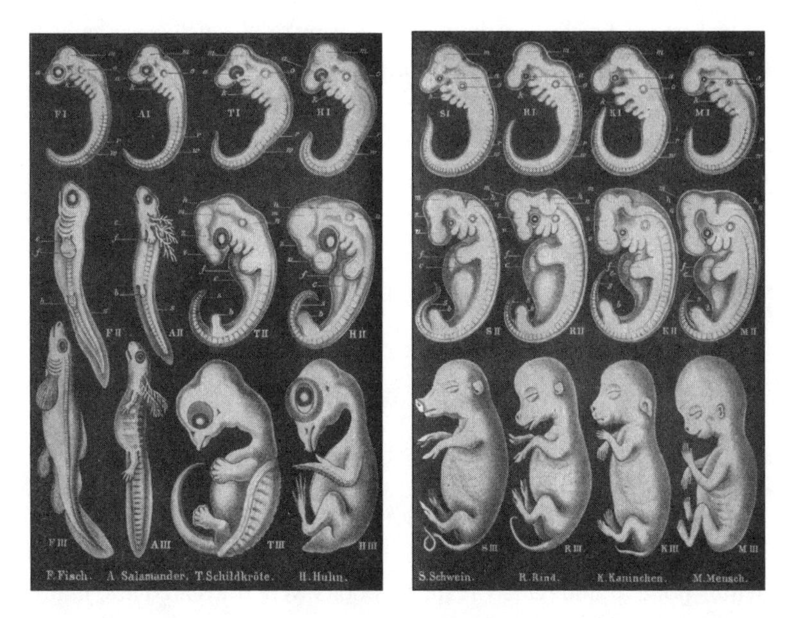

Figure 2.2. Embryo comparison by Ernst Haeckel. Haeckel presented the embryonic stages of different types of organisms to show how closely alike they are at the beginning and how they diverge. Though his overenthusiastic use of such diagrams drew criticism, the images have been used ever since in textbooks and elsewhere ostensibly to demonstrate ideas from evolution to creationism. From *Anthropogenie; oder, Entwickelungsgeschichte des Menschen, Keimes- und Stammesgeschichte* (Leipzig: W. Engelmann, 1877), Plates VI and VII, between pages 288 and 289 (explanation on p. 290).

since, there has been a great deal of scholarship, and much dispute, about exactly what Haeckel meant, on which evidence he was drawing, and the impact of his point of view.

Robert Richards has discussed in detail the episodes in which Haeckel was accused of fabricating the images he presented to demonstrate his ideas. As Richards clarifies, Haeckel used the same image for two different descriptions and people noticed. Haeckel responded that it did not matter for his purposes at hand—though of course he knew he should not have made such a mistake, but he had been too eager to hurry into print. Richards makes the case that, not surprisingly, the attacks hurt Haeckel personally as well as hurting the cause of evolution by creating uncertainty and confusion. In addition, because Haeckel had been so closely associated with connecting embryos and evolution, many embryologists decided to distance themselves from evolution altogether, which was a loss for both fields.

Striking paradoxes followed. Some biologists reviled Haeckel and completely rejected his ideas, and yet the public made his work wildly popular and widely read; his books were translated into many languages and reprinted often. Haeckel provided a big and clear interpretation, with beautiful images as supporting evidence, in a way that tied together lots of different phenomena and ideas. In fact, Haeckel's diagrams (those images of the comparisons of early developmental stages) have continued to appear over and over in textbooks to the present day. Sometimes, as Karen Wellner has shown, the caption in the textbook does not even indicate that the images came from Haeckel, and sometimes the authors even back off from discussing the evolutionary relationships that Haeckel had originally developed as demonstrations.[18] Instead, in the latter half of the twentieth century, his illustrations have increasingly been presented as showing the diversity of organisms and relations among them but not always as evolutionary relationships. This use suggests a sort of benign neglect of the original intent of the images.

Looking inside Embryos at Cells

One opponent who attacked Haeckel most vehemently with the intention of discrediting him was Wilhelm His, whose careful empirical work in the final quarter of the nineteenth century established the stages of normal human embryonic development. He emphasized the mechanics of development and completely rejected the broad-brushed evolutionary and theoretical approach that Haeckel took. In addition, His very publicly ridiculed Haeckel's sloppy work and labeled it as fraud. His was among those embryologists who saw the evolutionary emphasis of Darwin and Haeckel as directing attention from the real problems of biology, namely, understanding embryos and their development. His saw it as a waste to study embryos to interpret evolution. Instead, he and his embryological contemporaries focused attention on the details of comparative embryology in a way that generated a great deal of interest in embryos in their own right. He endorsed the observational and experimental approaches that gained ascendance around 1900 and after. Additionally, His brought together the study of embryos and the study of cells.

By the late nineteenth century, cell biologists (called cytologists at the time) were carrying out detailed studies of cells. What is inside, how do cells move, how do they divide, how do they work together to make up a complex whole? A few leaders mapped out details of the structure, function, and development of cells. In Germany, Oscar and Richard Hertwig asked about the significance of the internal structure of cells and examined the intersection of what they could see and the general, theoretical interpretations of what that meant about the nature of life and of development.[19] Oscar Hertwig's volume on cells and tissues served as a textbook that summarized existing knowledge for the German-speaking world of researchers.

In the United States, Edmund Beecher Wilson's works in 1895 and 1896 became the classic English-language discussions of the role of the cell.[20] Wilson had been one of the first to study at the new Johns

Hopkins University biology department, the formation of which had followed the inspiration of leading graduate educational institutions in Germany and England. Wilson came from a farm background in Ohio, and the more he learned about biology, the more he became fascinated by cells and developing organisms. After writing the first major introductory textbook on general biology with his fellow Johns Hopkins graduate William Sedgwick, Wilson turned to a detailed study of cells.[21] Though his first jobs after receiving his doctorate included a series of short stays at Williams College, then the Massachusetts Institute of Technology, and then the excellent women's college Bryn Mawr, in 1891 he moved to Columbia University in New York where he remained for the rest of his career.

In New York, Wilson partnered with a photographer to record exactly what happens in the earliest stages of development, the early cell divisions. His first studies looked at the polychaete *Nereis* worms, which he collected at the Marine Biological Laboratory (MBL) in Woods Hole, Massachusetts. Wilson must have gone out on the dock of the Eel Pond to the laboratory, shone a light on the water to mimic the full moon, and caused the worms to rise to the surface toward the light and begin their mating dance. Wilson would have gathered them up in his collecting net, taken the jar back to the laboratory, collected the fertilized eggs, and put them under the microscope. He would keep watch throughout the night to see every cell division and then observe and record every step of the epigenetic process.[22] The fertilized egg cells that Wilson watched so closely began dividing, gradually into two cells, then four, then eight and so on. As they divided, the nucleus did a wonderful dance, complete with complex spindle fibers, asters, and other apparatuses to help the division along.

Wilson saw all this, but not with unaided eyes. The microscope was absolutely essential for the kind of work he was doing, and it took experience and careful attention to detail both in the observations and in the preparation of the embryos. Although it is possible to look through a microscope and see some features of the dividing and developing embryos, the type of details that interested Wilson depended on meticulous preparation.

Collecting, watching, and even photographing were just part of the process. Because researchers had no method to see directly inside the cells without killing them, they needed a series of steps to make effective observations. This remains one of the great ironies of early biology: to study the details of life, one quite often has to kill it. Just as William Harvey had to perform vivisections on living animals to observe their blood pumping, Wilson had to kill the cells to see their inner structures. Like other cytologists at the time, he used a series of techniques to fix the cells with a hardening agent so that they would remain in the same position. Then they were preserved, so that they would last longer. They then had to be sliced open, which usually involved using a microtome to make very precise, regular sections. Typically the researcher then would stain them, experimenting with different dyes to make the structures clearer. These techniques would have revealed some of the underlying detail of the cells so that the observer could now see the complex structures within. Today, researchers use a variety of dyes and computer enhancement techniques to reveal otherwise invisible structures, so the process extends what we can see.

In later works, Wilson represented in diagrams or line drawings what he thought was important about what he saw. For his *Atlas of Fertilization and Karyokinesis of the Ovum* in 1895, he photographed the details of cell division so that everyone else could also observe precisely what he had been seeing (Figure 2.3). The photographs showed in detail what occurs in the cytoplasm and the nucleus as well as all the other small structures that were not very well understood at the time. Such photography was not easy, and capturing the details of cell division can remain a challenge even today with improved technology. Today, photography and even capturing observations on video or with increasingly sophisticated digital technologies has become a

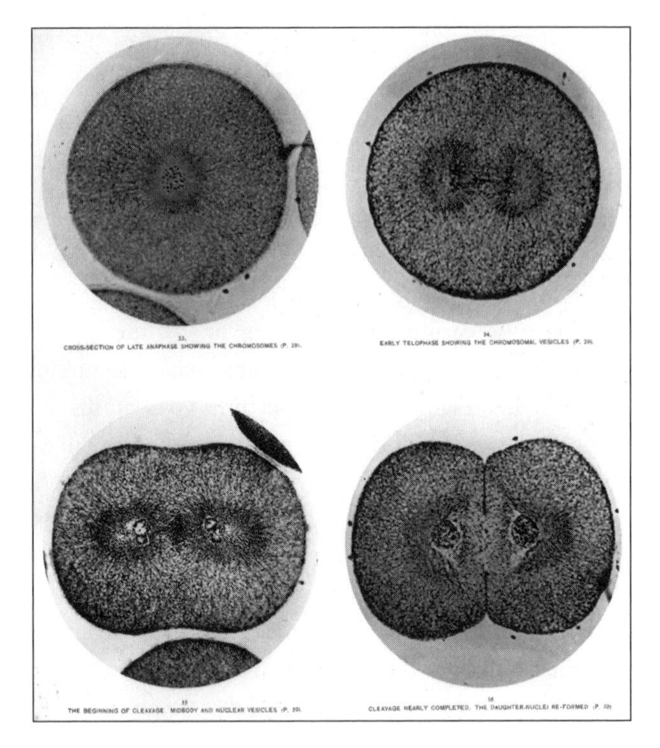

Figure 2.3. Details of cell division from Edmund Beecher Wilson's *Atlas of Fertilization.* In these meticulous preparations of *Nereis* worms, we can see spindle fibers, asters, and chromosome division. Wilson worked with photographer Edward Leaming to produce the book's ten plates, which became the first published photographs of cell division. Wilson's key to the illustration: "Plate IX, phototype 33: Cross-section of the karyokinetic figure, showing the chromosomes. This specimen is of the same stage as the last, but viewed in cross-section (i.e., from the end of the spindle). In the section, thirty-eight chromosomes can be counted, but not more than thirty are in focus. Half of these are of paternal, half of maternal, origin. This view shows the important fact that the chromosomes do not surround the spindle, but extend through its entire diameter. There can, therefore, be no distinction between a 'central spindle' and a 'spindle-mantle.' If two kinds of spindle-fibres be present, they must be intermingled. But there is no evidence that the fibres are, in this case, of two kinds." From Edmund Beecher Wilson, *An Atlas of Fertilization and Karyokinesis of the Ovum* (New York: Macmillan, 1895), 28.

routine part of cell biology, but in Wilson's day, it was rare to take photographs and expensive to publish them; a researcher made a considerable commitment. It is worth reflecting on how often today we just snap a photograph without thinking much about the process or about what selections we make in choosing one image or video over another. Wilson thought through his selections very carefully to share his observations with his readers, and it is fascinating to see what he considered important enough to record.

In 1895, Wilson enlisted the photographic skills of Edward Leaming, Columbia University supported what must have been a highly expensive publication, and the large format *Atlas* appeared in print.[23] These photographs were magnificently detailed and show the steps of cell division—we can observe the way that mitotic apparatus works as the cells divide and as the chromosomes divide. Wilson gave us the first published photographs of chromosomal division. He assumed that every reader would see what he saw in the photos, so rather than extracting or abstracting the parts he considered most important, he presented the "complete facts" to the public through the images.

The next year, Wilson took a different approach when he published the first edition of the magnificent *The Cell in Development and Inheritance,* which included no photographs.[24] There Wilson presented in textbook form the best available summary of what he knew. He discussed the various interpretations offered for all aspects of cells in the early stages of organisms, which allowed much more nuanced discussions than the presentation of photographic "reality." Instead of photographs, he gave his readers diagrams and drawings. In effect, Wilson was not just presenting but actually re-presenting his subject matter, so the use of images in this book are worth careful study as well.

In fact, the content of all three editions of the book deserve close analysis, because we can see how knowledge accumulated over time and also how new discoveries led to reinterpretations of what had gone before. When the third edition appeared in 1925 as the *Cell in Development and Heredity,* the change in title reflected the changed understanding of how heredity works through genetics. This final edition was more than four times the length of the first (1,232 pages compared with the 371 pages of the first edition).[25] The third edition was considered required reading for generations of biologists, and most senior cell biologists can recall the first time they read it. In fact, Wilson is so revered in the field of cell biology that the American Society for Cell Biology awards the E. B. Wilson Medal as its highest scientific honor, and a recent recipient has acknowledged his gratitude by referring to Wilson's work as "monumental."[26]

Cell Lineage

Wilson was a leader among the group of researchers who spent their summers at the MBL collecting and studying cells, including their detailed and incredibly meticulous studies of cell lineage. That research meant patiently observing the lineage of every cell from the initial egg to the division into two, into four, and so on (Figure 2.4). These divisions do not always occur in a completely regular way, because some cells divide faster or more slowly than others, and daughter cells are not always of the same size. Cell lineage researchers wanted to capture all the differences. They also wanted to record the details for many different species and compare the results because they were not sure which details would prove important and which were just accidents of the particular situation.

The MBL's first director, Charles Otis Whitman (an American who had taught at the Tokyo Imperial University, then at Clark University in Massachusetts, then for the rest of his career at the University of Chicago), led the way in inspiring his graduate students and colleagues to take up different invertebrate species and trace the lineages and the fates of each of the cells during development. Whitman studied the freshwater leech *Clepsine,* Wilson studied the polychaete worm *Nereis,* Edmund Grant Conklin studied the slipper snail *Crepidula,* and Frank Rattray Lillie studied the freshwater mussel *Unio*. Others took up other organisms, alongside these leaders; Lillie became the first assistant director and then the second MBL director, which meant that cell lineage remained a part of the MBL's central research from the 1890s into the early twentieth century. The record of lectures presented at the MBL through the 1890s reflects this strong interest in the role of cells and cell lineage in development.[27]

Cell lineage research showed a set of shared behaviors in the way that different kinds of cells in different organisms divide, and it also revealed the significant differences in how they respond to changes in their environment. Tracing every single cell and its path through the early developmental stages (for as long as it was possible to see and distinguish the cells) revealed patterns that are not visible by looking at just one point in time. Descriptive cell lineage studies provided an emerging picture of the process of early development and differentiation.

Researchers then went beyond observations and began experimental manipulations, such as compressing the embryo between two glass slides to see what happens as a result. Do the cells keep dividing in the same patterns but now spread out flat? Or does the sequence of cell divisions actually change in response to the different relationships among the cells that result from the pressure? Similarly, researchers asked what happens if they pressed the embryo between two slides and then rotated the whole slide upside down to change the gravitational field. Would heat or cold make a difference, or perhaps changes in light? What variables mattered?

Cell lineage work began as observation of normal conditions, but quickly shifted to a more experimental approach. The researchers learned, as happens so often in science, that . . . it depends. For some organisms, under some conditions, the cells would smush out flat and then recover their normal patterns right away; others did

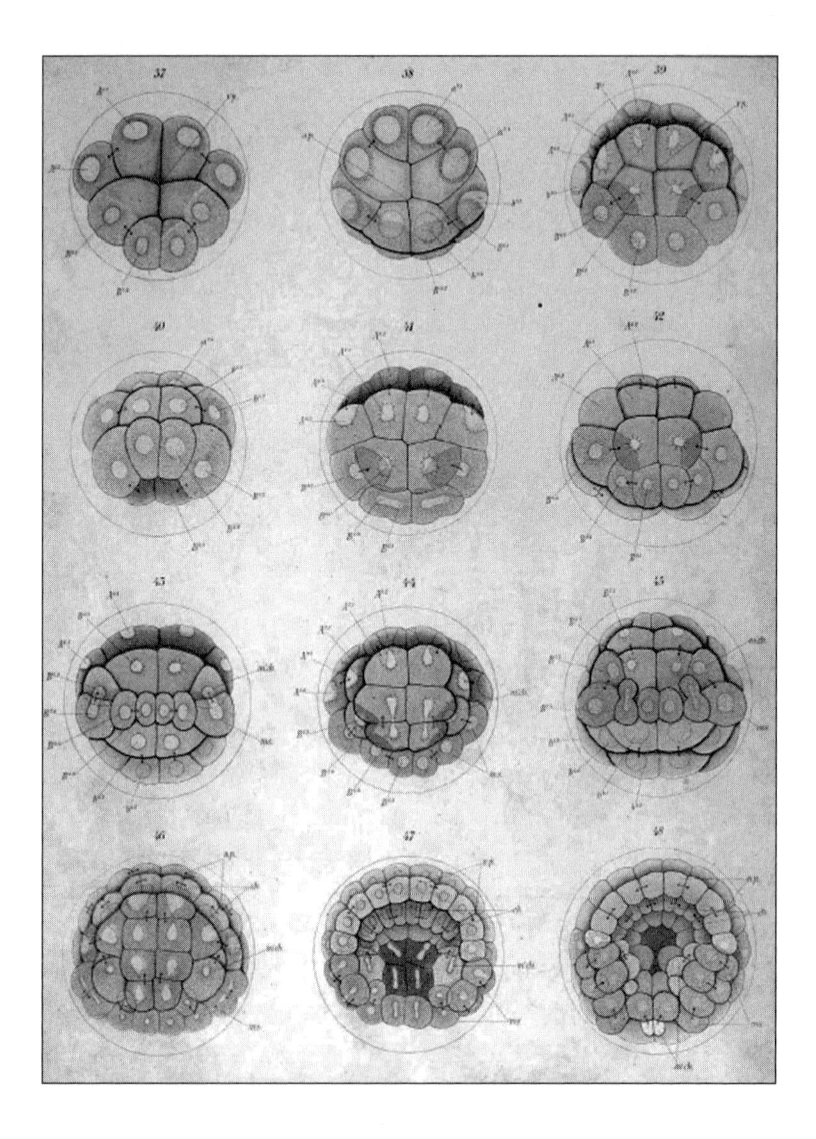
37
38
39
40
41
42
43
44
45
46
47
48

Figure 2.4. Cell lineage work by Edwin Grant Conklin. Conklin joined Wilson and others in tracing details of cell lineage in a variety of marine organisms. His report on the Ascidian egg was a beautiful summary of meticulous, detailed comparative work, following each cell through every cell division to see what changes occurred and how they varied across different types of organisms. Shown here is the final plate of the series. "The Organization and Cell-Lineage of the Ascidian Egg," *Journal of the Academy of Natural Sciences of Philadelphia* 13 (1905): 1–119.

Conklin's key to the illustration

37. Sixteen-cell stage viewed from vegetal pole.
38. Sixteen-cell stage viewed from the animal pole, yellow protoplasm around the nuclei.
39. Twenty-two-cell stage viewed from the vegetal pole; 4 mesoderm cells (yellow), 10 endoderm, chorda and neural plate cells (gray), and 8 ectoderm cells (clear).
40. Same stage viewed from the posterior pole.
41. Egg passing into the 32-cell stage, postero-dorsal (vegetal pole) view.
42. Thirty-two-cell stage, postero-dorsal view.
43. Forty-four-cell stage, posterior view, showing separation of mesenchyme (m'ch) from muscle cells (ms.).
44. Same stage, dorsal view, showing subdivision of endoderm cells.
45. Similar stage, posterior view, showing separation of another mesenchyme cell from a muscle cell.
46. Seventy-four-cell stage, dorsal view, showing division of 4 chorda and 4 neural plate cells; there are 10 mesenchyme and 6 muscle cells, besides 10 endoderm cells.
47. One-hundred-and-sixteen-cell stage showing the beginning of gastrulation and also the neural plate chorda, muscle, and mesenchyme cells.
48. Slightly older stage showing advancing gastrulation with inrolling of cells at edge of blastopore.

not. Some would change in response to the altered gravitational field; others would not. The challenge, then, was to discover what makes the changes occur. Which results are the same time after time after time for every embryo confronted with the same circumstances? Which ones vary in ways that suggest that the embryo has the ability to respond to changes and adapt?

Cell lineage studies led to many new questions, and by the early twentieth century those new questions and experimental approaches led to the field of experimental embryology. The meticulous, incredibly time-consuming cell lineage studies largely gave way to other approaches—that is, until new technologies allowed researchers to study the lineage of cells in the nematode worm *Caenorhabditis elegans* in the late twentieth century for new reasons and with new results. By then, having what was called a model organism with a small number of cells made it possible to trace each cell, to compare patterns and relationships, and to begin interpreting which genes were driving the cell divisions and differentiations. Sydney Brenner, H. Robert Horvitz, and John E. Sulston shared a Nobel Prize for their meticulous study and for showing how specific genes regulate organ development and programmed cell death.[28]

Normal Stages of Development

The patterns revealed by cell lineage studies complemented the patterns observed in establishing standard tables of normal stages of embryonic development. There were several different motivations for doing the careful work required to lay out the stages in detail, inspired in part by an interest in evolution and the resulting need to demonstrate relationships among species.[29] Medical interests also helped drive the collection of human embryos, just as excitement about understanding biological processes and patterns motivated the collection of diverse species. If researchers were to understand what variables make a difference in development—that is, whether changes in temperature or pressure or gravitational direction or light or such things matter—then they needed to know far more about what is the normal set of patterns and processes for embryos. In particular, it was valuable to map out what came to be called the normal stages of embryonic development.

For example, what are the normal stages of frog development? Frogs were playing an important role in understanding development and had long been a martyr to biology.[30] In the eighteenth century, for example, Luigi Galvani had hung frog legs over his patio, applied an electrical current, and watched as they kicked—apparently just as they do in life. As the story goes, he eagerly waited for a lightning strike to test the effects of "natural" electricity as well. Physiologists especially found frogs an easy, relatively cheap, available source of experimental subjects on which to test all manner of questions about function. As we saw with von Baer, frog eggs also held particular appeal because they are so easy to collect and to see. By the early twentieth century, a number of embryologists had taken up studying frog eggs, both observing them directly and experimenting on them to see what variables affect development.

Ross Granville Harrison established the normal stages of frog and salamander development, and he handed out copies of the normal plates to all his graduate students so that they all had the same starting point (Figure 2.5). This aspect of controlling and being clear about what is normal in order to study and interpret deviations and thereby to get at causes is a central feature of experimental biology. Students were expected to experiment by moving around cells or pieces of tissue from one location or even one frog to another. They were then expected to compare their results to the normal table. The table of normal stages showed not just what was different, but also whether the timing of developmental stages changed because of the experimental manipulations.

In addition to learning about development in frogs and other accessible species, there was tremendous interest in the early development

Figure 2.5. Ross Harrison's normal stages of *Amblystoma* (later renamed *Ambystoma*). Harrison provided all his students with a copy of this illustration of the normal stages in frog development. He intended it to serve as a standard template against which his students could compare any divergences that resulted from experimental manipulations. (Illustration courtesy of Harrison's student John Shaver, who gave it to the author.)

of human embryos in particular. Under normal conditions, human embryos remain invisible. In Germany, Wilhelm His collected every human embryo (which at first meant only those beyond the eight-week or technical embryo stage, and thus technically fetuses) that he could find as well as nonhuman embryos, and he lined them up to document each stage. His worked with Franz Keibel, who in turn worked with the American Franklin Paine Mall. Early in the twentieth century, their collection moved to the United States to become the Carnegie Collection, under Mall's direction.[31] The Carnegie Embryo Collection, including what are known as the Carnegie Stages, presented examples of each developmental step (Figure 2.6). When they are all lined up together, they look so logical and neat that it is easy to forget how much effort went into creating and documenting the collection.

Although the later stage fetuses came from natural or induced abortions, the very earliest and smallest stages were found largely by accident—often through autopsies of dead women who turned out to have been pregnant. It took an army of physicians to collect all the specimens, and they held the conviction that it was important to preserve those specimens and then send them to the accumulating collection housed first at the Johns Hopkins University and then later the Armed Forces Institute of Pathology at the Walter Reed Medical Center in Washington, D.C., until that facility closed in 2011. In fact, the state of Maryland's public health department called for physicians to donate any embryos, with the expectation that they would be named after the doctor donating them and with the understanding that they had thereby contributed to advancing scientific knowledge and presumably also public health.

Wilhelm His, who started the human embryo collection, also made other important contributions to seeing embryos in more mechanistic terms. He decisively rejected any vital factors in shaping embryos and insisted that epigenetic development from one step to the next was driven by mechanical forces. It was the folding, unfolding, and

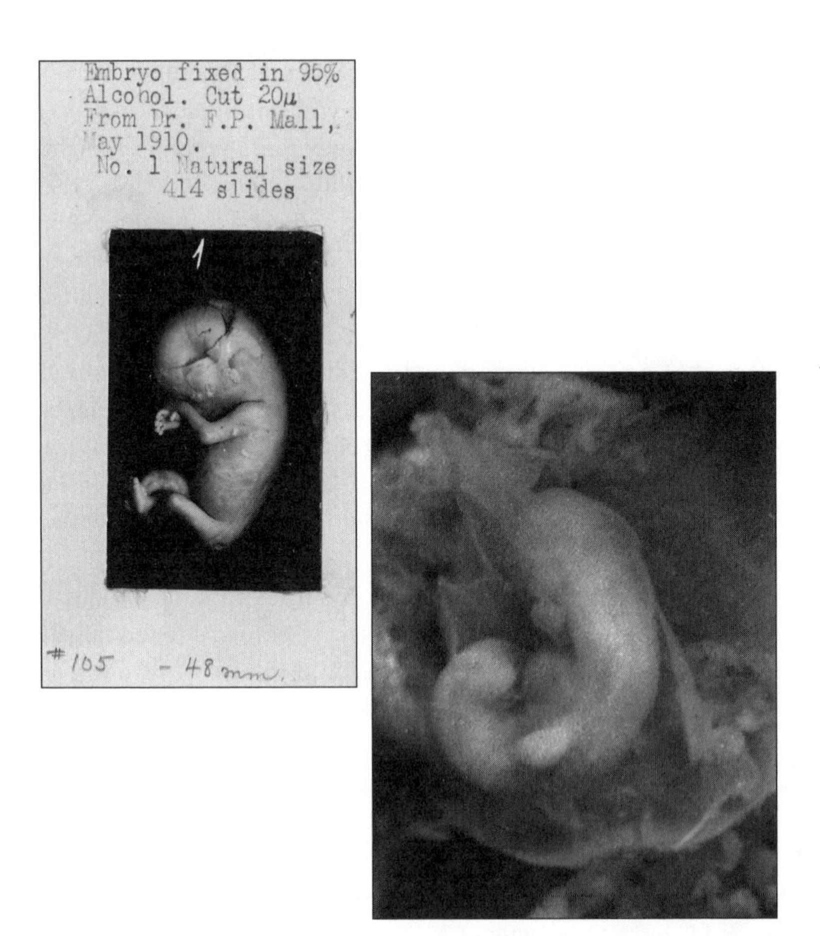

Figure 2.6. Examples of Carnegie specimens. The Carnegie Institution of Washington, Department of Embryology, collected human embryos and fetuses under directors Franklin Paine Mall and George Streeter. The fetus specimen (upper left) was collected by Mall in 1910; it was used in drawings before being sliced into serial sections to reveal the specimen's insides. The embryo specimen (lower right), aged about two months, was collected in 1895, and the collection retains documentation of how it was used. The collection's records indicate that both specimens were later cremated. Used with permission of the Human Developmental Anatomy Center, National Museum of Health and Medicine. My thanks to Adrianne Noe and Elizabeth Lockett for their help.

refolding of material in different ways that caused embryos to develop from unformed clusters of cells into formed organisms. For His, every organism starts out as a cluster of material that then undergoes physical foldings and structural changes. In his 1874 book about the problems of form and physiological processes of development of individual organisms, His used a series of essays to illuminate the motions of cells and especially the folding of germ layers (Figure 2.7).[32]

Germ layers had been hinted at by Wolff and introduced by von Baer, who described the way that as the number of cells increases, they fold into layers that seem to be consistent across different individuals and even different species. Von Baer initially thought there were two layers, which then seemed to divide further. Others such as Christian Pander and Robert Remak continued to explore the layers, and Pander and His carefully sorted out that there are typically three layers. He saw what he thought of as "organ-forming germ layers" ("organbildenden Keimbezirke") and structured his set of essays around interpreting how these move and change during development to bring differentiation and specialization to the embryo. In his observations, he discovered a number of developing structures, including the endothelium, which he named as such. The endothelium is an extremely important layer that covers the inside of the blood vessels. We know today that the endothelium includes stem cells and helps make possible normal vascular function and wound repair as well as sustaining life. His's approach made it possible to see such structures.

His emphasized most emphatically, and then other embryologists strongly reinforced the idea, that researchers interested in embryos should look closely at embryos and the details of development, and not assume that they can imagine what is happening based on their underlying metaphysical convictions about how life must work. His invented an improved microtome, a device for taking the preserved embryo specimens and slicing them with a very sharp knife into

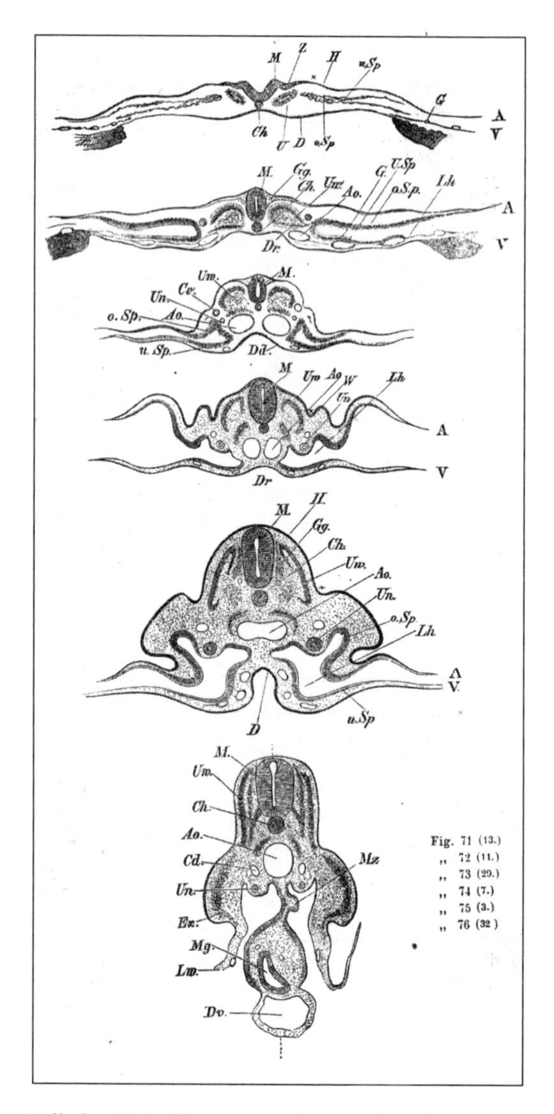

Figure 2.7. Wilhelm His's illustration of organ-forming germ regions. In 1874, His presented his theory and evidence to explain the process of embryonic development; this was a mechanical account that emphasized the folding and unfolding of parts to shape the embryo into an organized organism. Wilhelm His, *Unsere Körperform und das physiologische Problem ihrer Enstehung: Briefe an einen Befreundeten Naturforscher* (Leipzig: F. C. W. Vogel, 1874), 84.

serial sections. The microtome works something like a meat slicer; when a cell or embryo is run through the microtome, very thin slices result. When the slices are lined up, one can "see inside" the whole embryo and observe how the structures are put together and how they change over time. His's microtome, together with improved microscopes at the end of the nineteenth century, made it possible to view a great deal more detail of developmental stages than before. His helped make embryology into a serious empirical science with his meticulous observations and imaging techniques.

Soon, however, it became clear that no matter how closely they looked, embryologists could still not observe many details inside developing organisms or most of the process. Seeing more and seeing in more detail called for new approaches, and experimental embryology brought new tools and a new way of thinking about development. Perhaps, the experimental embryologists thought, they could actually see better by manipulating and controlling normal development in order to compare normal with experimental conditions. In the 1890s and after, experimental embryology grew as a field and began to yield considerable understanding about how embryos develop.

3

Experimental Embryos in the Laboratory

Experimentation made possible an additional source of information and interpretation to take the observer beyond what he (the researchers were still mostly male well into the twentieth century) could normally see. The distinguished historian of science Garland Allen has addressed the move toward explicitly experimental approaches to biology in his textbook *Life Science in the Twentieth Century.*[1] He gives us a look not just at the separate specialties of the life sciences, but also at the idea (and ideal) of life science as a whole with many parts that work together. Furthermore, Allen puts forth the interpretation that the new experimental science involved a "revolt from morphology." That is, he viewed researchers as having become dissatisfied with the limited answers yielded by merely observing and theorizing. They found the speculative thinking of evolutionary theorists especially inappropriate to their ambitions to ground life science on foundations as solid as those of the physical sciences. Allen sees this as a revolt from speculative and descriptive thinking around 1900 toward an eager desire for change and for modernization.

Several graduate students at the time of Allen's *Life Sciences* publication (basically one generation later) did not see the changes in the field of biology in quite the same way. Our counter was that it had not been so much a revolt from morphology as an extension of its study (or the study of form) with additional experimental approaches.[2] In part, this is because we were each focusing on different aspects of biology; it was not so much a rejection of the old but an embracing of the new experimental methods alongside the older, traditional approaches and ideas.

Some scientific change is gradual and incremental; some is more purposeful, a radical and intentional break from tradition. This issue is worth continued reflection because scientists often claim to be doing something radically new or "revolutionary" or "transformative" (as various generations have labeled major changes), and they are often rewarded for the apparent newness of their work. Yet typically they are building in interesting ways on work that came before; having scientists acknowledge their predecessors more often and more carefully might well be good for science, for presenting the discovery process more accurately. This book presents the evolving knowledge about embryos in a way that shows how new ideas build on previous ideas. Even the most amazing new discoveries of the moment are revolutionary only in the sense of involving a larger jump in knowledge than before, or a shift in the understanding and meanings of the object being studied. Transformations may occur in ideas and interpretations, but they are always drawing on previous knowledge.

The question at hand in this chapter is whether biologists at the turn of the twentieth century saw themselves as intentionally different from their predecessors—and the answer is surely yes, they saw themselves as experimental in a new way. But the question remains as to whether they saw this as a revolt that was a break from their predecessors—and what that might mean for today's emphasis on "transformative science." Perhaps historical cases can shed light on our understanding of what makes science advance today. Wilhelm Roux, for example, most definitely saw himself as a revolutionary.

Half-Embryos of Roux and Driesch

Experimentation drew on a number of methods, all intended to provide ways of getting information about what was inside the developing embryo and about the patterns and processes as well as the causes of development. The move toward experimentation was formally articulated in 1894 with Wilhelm Roux's manifesto for the journal he had founded and edited, *Archiv für Entwickelungsmechanik*. This publication, known as *Roux's Archiv,* was later renamed *Development Genes Evolution.* In his journal, Roux called for the study of the mechanics of development, which he acknowledged that some saw as a small specialty field but that he viewed as the basis for understanding core issues in biology. Roux's 1895 introduction to the journal was widely read, discussed, and taken as a manifesto for a new way of doing science. An English translation was even published as part of the Marine Biological Laboratory (MBL) *Biological Lectures* and was cited often—both positively and in opposition to Roux's particular interpretations.[3]

Roux explained that this new field should seek to understand the causes of organic form in mechanical terms. Just as physics and chemistry can be explained in terms of movement of material, so can biology, including development. Formative forces might exist, but they are purely mechanical and leave no room for old-fashioned ideas of special vital forces or factors, Roux insisted. Out of the apparent complexity of life, development nonetheless consists of simple processes that combine to bring an organized system out of the parts. Roux's emphasis on experimental methods and materialistic interpretations found agreement among his contemporaries, even among those who did not share his theoretical interpretations. The end of the nineteenth century brought a turn to the sort of mechanistic and experimental embryology that Roux advocated.

What followed was a new wave of debate about preformation and epigenesis. Roux's approach suggested a kind of preformation or predeterminism. For Roux, the nucleus and the hereditary units seemed to carry the determinants for the developmental processes that followed. This suggested that cutting off parts would lead to damaged embryos, for example. An alternative view by Hans Driesch suggested an ability of the organism to maintain its integrity and autonomy by adapting to changing conditions or through its capacity to self-regulate. Driesch discussed his ideas in terms of what he called regulation, and he worked to sort out what it was about the individual embryo that caused it to become an organized individual whole organism, even when it had cells missing or parts damaged.

Driesch's interpretation led him back to Aristotle, and to the idea of an entelechy for each organism. Entelechy was an innate vital principle that helped make sure that the organism became actualized as the right kind of thing. Driesch eventually decided that biologists did not know enough to understand embryonic processes, and he turned to philosophy, even becoming a professor of philosophy and largely giving up biological research. His idea of an entelechy embraced a kind of vitalistic thinking that other biologists and philosophers of science such as Rudolf Carnap did not accept. Some of them, like Carnap, found Driesch an original thinker but were puzzled by his willingness to give up what they saw as proper scientific thinking. Others did not accept his approach at all, and saw Driesch as too distracted by metaphysics while failing to keep the actual physical embryos and their development at the center of his ideas. These biologists kept working to discover the causes that direct such complex processes in such reliable ways.

Roux's own work, as the distinguished neuroembryologist-turned-historian Viktor Hamburger noted later, was itself limited.[4] Hamburger pointed out that Roux lacked much of a visual feel for the embryos he studied. Most others who studied embryos included images of what they saw, but Roux included very few illustrations with his publications. Instead, he was inclined to draw strong conclusions from the available evidence he had available, even when it was limited. We might also add that Roux had a rather high opinion of himself and his research, and he would cite his own work extensively. In fact, half the references in his manifesto were to his own publications. He also liked to rewrite the papers of the authors submitting to his journal, and he often inserted references to his own work in other authors' papers. As an editor, Roux did not promote the idea of peer review, which was already becoming a standard for the best scholarly publications. Instead, he happily made all the editorial decisions himself. In the 1890s, Roux's journal was nonetheless the best place to publish developmental papers, so authors put up with Roux until they founded other journals to provide additional outlets early in the twentieth century. For all these reasons, as well as his particular experiments, Roux had a major impact on experimental embryology. He could not be ignored.

In 1888, Roux reported his experiments with frog embryos that appeared to provide support for his own special brand of determinism.[5] Roux explained that he had a way to address the question of what impact each cell division has on the development that follows. Normally, an egg cell is fertilized and begins dividing, first into two cells and then into four and so on. Roux hypothesized that each cell division makes each of the resulting cells different in an important way. That is, the cells differentiate with each division, and he believed that this was because the chromosomes divide in ways that make each cell qualitatively different from each other. The result, he thought, was a sort of mosaic—in which the separate mosaic tiles are each different and can be arranged in different ways to make patterns that differ in details. But once a mosaic tile is, say, green or red, then it will not revert to an earlier uncolored state. Roux hypothesized that this happens with cells after they divide.

This led him to predict that when the first cell divides into two, the two parts should be different enough that if he could kill one of the cells, the other would develop as a half-embryo. This is what he predicted, and he performed the experiment with frog embryos. He could not actually separate the two cells without killing the whole thing, so he did the next best thing and poked one of the cells with a hot needle for long enough that it stopped developing and apparently died. The other cell proceeded to develop, and in all ways seemed to behave normally—it seemed to be operating as half an embryo. Roux concluded that his hypothesis about differentiation was confirmed: development follows a mosaic pattern (Figure 3.1).

If you read the last two paragraphs quickly and know something about biology today, you might have been tempted to think that Roux was right. But that is because when we read today, we tend to put in our own assumptions as if they were true in the past as well. It is true that Roux was focusing on the importance of chromosomes for development, as we do today. But he was not saying that they are passed on as a whole to every cell. Instead, he was saying that they are divided up with each cell division, which is what produces that mosaic. This seemed to make sense because it provided a neat explanation for how cells become differentiated. Otherwise, as Frank Lillie and others have pointed out in what has been called "Lillie's paradox," there is a huge biological question about how chromosomes and cells that are essentially the same can become so different and can make up the highly differentiated, complex, interacting system of parts. Roux solved that problem by separating the chromosomal materials into the different cells to make them become different. We know today that although this is a clever theory, it is not the way chromosomes actually work. In fact, each cell gets the same complement of chromosomes, and it is other factors that cause differentiation.[6]

Roux studied frogs, but he was confident that the same thing would happen with other species and under other conditions. His fellow researcher Hans Driesch decided to try the experiment with a different organism, in his case using sea urchin eggs because it was known that shaking the urchin eggs would cause the cells to separate completely. Driesch reasoned that it would be a better experiment to

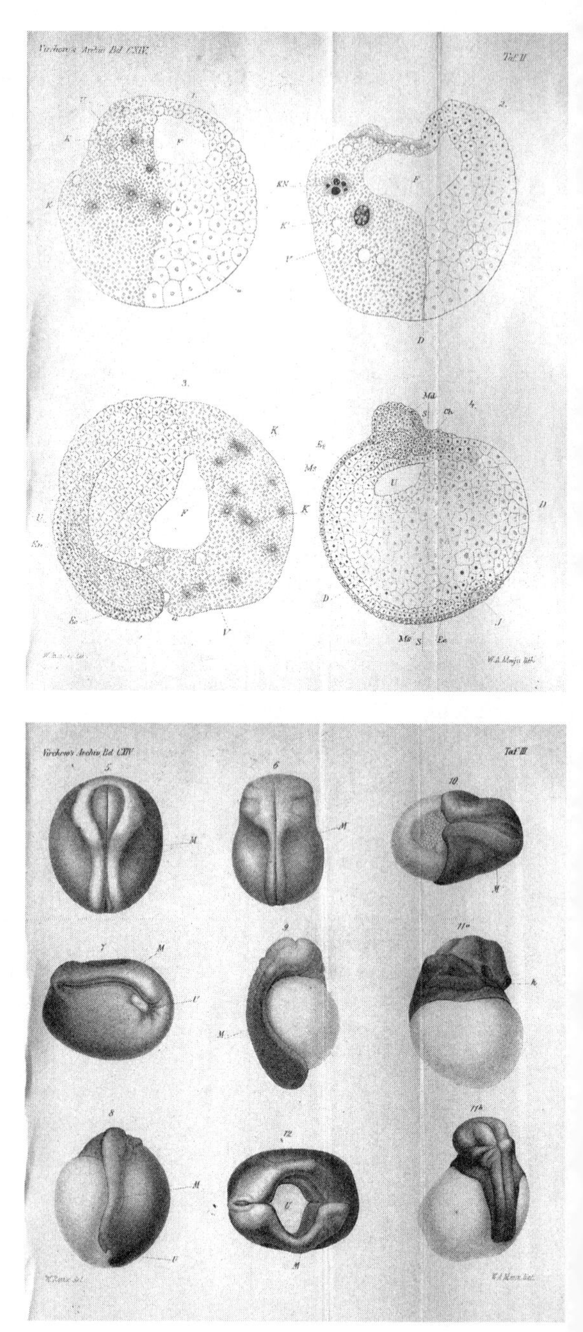

Figure 3.1. Wilhelm Roux's half-embryo experiments. After Roux punctured frog eggs with a hot needle, he observed that only half the frog embryo developed normally and the other half died—which was in line with his predictions. For Roux, this proved his predeterminist mosaic interpretation of development. Wilhelm Roux, "Beiträge zur Entwickelungs-mechanik des Embryo, No. 5: Über die künstliche Hervorbrin-gung halber Embryonen durch Zerstörung einer der beiden ersten Furchungskugeln, sowie über die Nachentwickelung (Postgenera-tion) der fehlenden Körperhälfte," *Virchow's Archiv für Patholo-gisches Anatomie und Physiologie und klinische Medizin,* 114 (1888): 113–153.

use cells that were actually separated and not just killed with a hot needle and still sitting there attached as seemingly inert material. Driesch performed his experiment by shaking the two cells and noted that he went to sleep while the cells were going through their early divisions. He reported that he fully expected to wake up the next morning to two separated cells that were developing nicely as half-embryos, as in Roux's report but far more definitively.

But that was not what happened. Instead, he found two fully formed embryos that were a bit smaller than usual (Figure 3.2). Driesch concluded that something about the fact that the one cell was now by itself caused it to regulate as if it were a whole, separate, and individual organism, even though it only had half the material it had started with. It also should have only half the chromosomes, according to Roux's interpretation. Driesch then showed that the same thing happened with four cells, or even eight. In each case, the separated cells became a whole embryonic form. For Driesch, this meant that each cell at that point was truly "totipotent," meaning that it had the ability to become a "total" or whole organism by itself.[7] Roux's interpretation of chromosomal separation and early differentiation could not be correct.

The fact that Roux and Driesch reached different results stimulated lively discussion and caused others to jump in with new experiments. Surely Roux had not anticipated that a major provocation for work in experimental embryology would come from criticisms of his own experiments, but it worked that way. Roux and Driesch were engaged in addressing questions about the extent to which each cell becomes different from every other cell, and how this happens. Their research pointed to core questions about how differentiation occurs, and the extent and way in which it is essentially self-differentiating or, alternatively, driven by external forces and factors. Roux was confident enough that he made up some ad hoc interpretations that he called "auxiliary hypotheses." He suggested, for example, that each cell also has a special "reserve idioplasm" that kicks into effect when

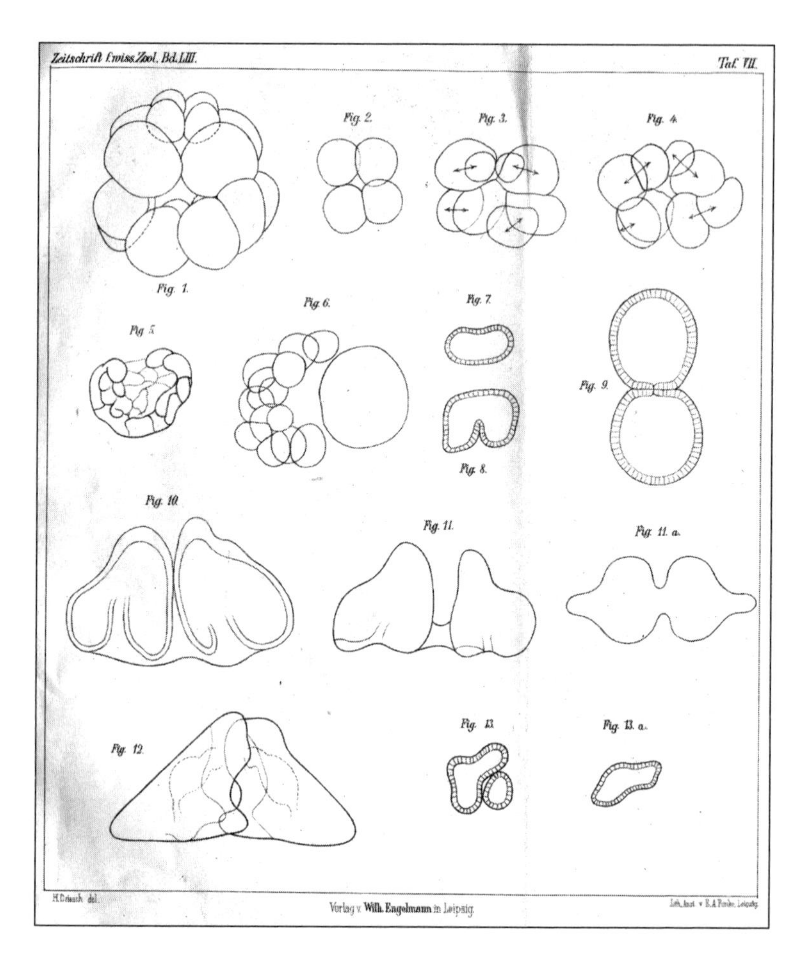

Figure 3.2. Hans Driesch's regulatory interpretation of sea urchin development. Where Roux saw half-embryos in frogs when one of two cells was destroyed, Driesch saw a process of regulation that allowed two separated cells each to develop a new form in sea urchins. Driesch's Figures 1–13 show a series of stages from 8- and 16-cell stages to the twin pluteus larval form. In Hans Driesch, "Enwicklungsmechanische Studien, I: Der Werth der beiden ersten Furchungszellen in der Echinodermentwicklung—Experimentelle Erzeugen von Theil- und Doppelbildung," *Zeitschrift für wissenschaftliche Zoologie,* 53 (1891): 160–178. Courtesy of the Marine Biological Laboratory, Rare Books Room.

there is an "injury," which happens when the cells are separated or one is killed. Very few others accepted this interpretation, but that left the question of how differentiation does occur.

Some researchers, including most famously Edmund Beecher Wilson and Thomas Hunt Morgan, took up the same experiments in other organisms, and they found that Driesch's regulation occurs in some cases while others are more like Roux's differentiations. What became clear is that experimental embryology offered a range of opportunities to explore what happens in development. Manipulating the embryo and cells to see what happens under different conditions brought new methods, augmented with new techniques and tools for carrying out the manipulations. The experimental approach brought new ways to gain new knowledge and to help sort out what actually goes on in those earlier developmental stages.

Boveri on the Nucleus and Chromosomes

A related set of core questions concerned the extent to which and the way in which different parts of each cell contribute to subsequent development and differentiation. In particular, what was the role of the nucleus versus the role of the cytoplasm? Here, the most important researcher was Theodor Boveri. Like Driesch, Boveri also worked with sea urchin eggs, which are plentiful and easy to study in a well-equipped and welcoming marine research laboratory such as the Stazione Zoologica in Naples where Driesch and Boveri each enjoyed performing their experiments right on the Bay of Naples.

Boveri realized that when they were shaken, not only did the urchin cells separate as Driesch had showed, but they could also break up into smaller fragments. He started with the conviction that under normal circumstances it is the chromosomes in the nucleus that carry hereditary information and that the chromosomes carry some sort of "determinants" that make them the important factor guiding development. He hypothesized that those pieces that contain whole nuclei have the capacity to develop fully, whereas those that contain none cannot develop at all. Fragments with partial nuclear material were thought to have varying capabilities. (Figure 3.3).

Historian and theoretical biologist Manfred Laubichler at Arizona State University and the distinguished Caltech biologist Eric Davidson have teamed up to study Boveri's hybridization experiments of 1889 in modern terms and from a historical perspective. They have shown that these beautifully controlled experiments remain especially impressive because of the theoretical interpretations they made possible. In his 1889 work, Boveri made hybrids by combining parts of cells from two different sea urchin species that were clearly distinguishable. He took a fragment from the cell of one species and introduced the nuclear, chromosomal material from another species. This type of recombination experiment allowed Boveri to examine the relative roles of the nucleus and cytoplasm in development by simply observing the development that followed. He concluded that the nucleus is what directs and determines development. This was a strong claim that was not immediately accepted by many, but it held rich possibilities and laid the groundwork for subsequent studies.[8]

Nucleus or cytoplasm? Chromosomal determinants or regulatory response? Internal or external factors? What drives development, and how can we best study it? Wilson, whose own work and the work of his students added considerably to the understanding of the role that chromosomes play in early cell division, captured a widely shared view at the end of his book *The Cell in Development and Inheritance* in 1896. In this volume, which he dedicated to Boveri, Wilson wrote, "I can only express my conviction that the magnitude of the problem of development, whether ontogenetic or phylogenetic, has been underestimated; and that the progress of science is retarded rather than advanced by a premature attack upon its ultimate problems. Yet the splendid achievements of cell-research in the past twenty years stand as the promise of its possibilities for the future, and we need set no limit to its advance."[9] Wilson could have written that statement today.

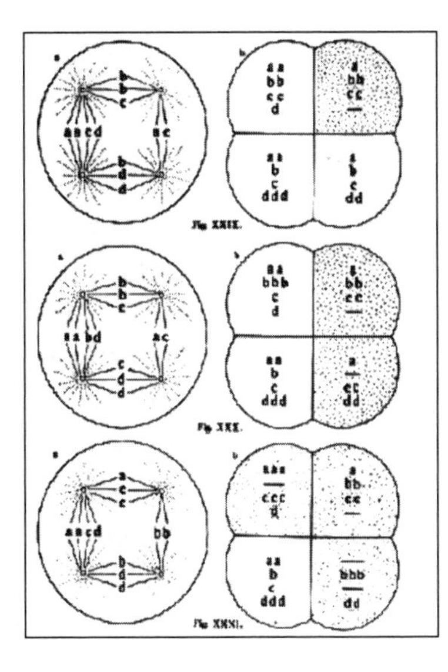

Figure 3.3. Theodor Boveri's model of chromosome distribution with cell division. Boveri performed many elegant experiments to demonstrate the pattern of chromosome distribution and chromosome roles in cell division. Images like this appear in several of his publications, most notably in his major paper "Zellen-studien, VI: Die Entwicklung dispermer Seeigeleier—Ein Beiträge zur Beifruchtungslehre und der Theorie des Kernes Jenaische," *Zeitschrift für Naturwissenschaft* 43 (1907): 1–292.

The past twenty years have raised so many great new questions and have provided so many insights into development. And, yet, as Wilson wisely noted, we should not rush to conclusions or underestimate how challenging it is to gain a full understanding of all the developmental processes and patterns. Research such as Jacques Loeb's showed that surprises could open new doors to understanding.

Other experimental research in the late nineteenth and early twentieth centuries reinforced Wilson's view that the situation was tremendously exciting but there were perhaps more questions than answers. In addition to the work that Boveri inspired on the importance of chromosomes and the nucleus, four areas deserve additional attention: Jacques Loeb's work on parthenogenesis, the debates about fertilization, Thomas Hunt Morgan's study of regeneration, and the experiments with transplantation of tissue and cells, especially those by Hans Spemann and Ross Granville Harrison. Interestingly, given their importance around 1900, it is these lines of work that have come together again at the beginning of the twenty-first century and are raising new questions and possibilities in the context of what has been called regenerative medicine.

Loeb's Parthenogenesis

Like Wilson, Jacques Loeb spent his summers doing research at the MBL in Woods Hole. Like Wilson, he had taught at Bryn Mawr before going on to a larger university, in his case the University of Chicago. And like Wilson, Loeb was interested in development. Unlike Wilson, however, Loeb cared more about the physiological factors that cause change and less about the structure and behavior of cells in particular. He wanted to know about the dynamic action of development.

Therefore, Loeb was excited when he discovered somewhat by accident that sea urchin eggs could develop parthenogenetically under some conditions. This means that sea urchin eggs, which under normal circumstances would be fertilized and begin dividing and differentiating in the same way as other sexually reproducing organisms, could start differentiating without fertilization. This could be caused by such environmental conditions as changing the salt water content or physically pricking the egg to begin the division process without it having been fertilized. Loeb's discovery was hailed by the public press as a sort of virgin birth, in which the egg contribution from the mother could carry out the whole job of starting development by itself, without the normal male contribution. Newspapers reported that a scientist had discovered "the secret of life," the "origin of individuality," and other exaggerated claims.

Obviously, the research results did raise many new scientific questions about the relative contributions of each parent as well as about the nature and role of fertilization under normal circumstances. The experiment revealed that the embryo is highly flexible in ways that added support for Driesch's interpretation, arguing for the embryo's ability to regulate under changes in cell division. Embryos could respond to changed environmental conditions and nonetheless develop in what seemed to be normal circumstances. As the late Rutgers University historian Philip Pauly explained in his close look at Loeb's philosophy and experimental work, Loeb endorsed an engineering conception of life.[10] Loeb was, in many ways, one of the first to embrace a version of the constructed embryo that became more common in the late twentieth and early twenty-first centuries, and his search for a mechanistic approach will reappear as a central theme in a later chapter in the context of efforts toward engineering organisms.

Fertilization

Another topic with many open questions concerned fertilization, and here Loeb also played a central role. Egg and sperm come together, a process that had been well-studied since the middle of the nineteenth century. The process sounds simple enough—but just how do two cells become one? Assuming that both the sperm and egg cells bring their chromosomes to the union, how do the chromosomes avoid doubling in number every time? And how does the process work? Is it a purely mechanical process, with two cells coming together and combining material, or is it a chemical process between the two cells?

Competing theories were intensely debated in the early twentieth century. In part, this debate played out at the MBL, where Frank Lillie served as the director and Ernest Everett Just spent time as a researcher. Lillie and Just developed one theory, which emphasized the interactions of the egg with its environment. Loeb offered another, competing theory that emphasized factors internal to the egg itself. The debates became heated, and they raised additional kinds of social as well as scientific questions because Just was the first African American to spend a significant amount of time at the MBL and to become a well-known American developmental biologist.[11]

Just studied with Lillie at the University of Chicago, and he began by embracing a theory very much like Lillie's. Just and Lillie saw the egg as a chemically active agent in fertilization, secreting a substance that they called "fertilizing" to attract the sperm to the egg to initiate the fertilization process. In contrast, Loeb emphasized that a particular substance, or cytolysin, breaks down the surface of the egg and allows the sperm to enter under normal conditions of fertilization. His theory could also explain how the egg could begin to divide even without a sperm, as in the case of artificial parthenogenesis, as long as the same substance caused the breakdown of the cell surface. Later interpretations would lie somewhere in between these two alternatives and bring in additional factors to explain the complex and fascinating process of fertilization and chromosomal division.

Morgan's Regeneration

Another approach that emphasized the adaptability of embryos under changing circumstances involved regeneration research. We have already seen that some researchers such as Trembley in the eighteenth century had been curious about hydra and their ability to respond to having their heads chopped off by regenerating new heads. In 1901, regeneration was a lively topic, and Thomas Hunt Morgan gave a series of lectures on the topic at Columbia University, where he had moved in 1904 after teaching at Bryn Mawr College. These lectures eventually became one of his many books, and he began with an outstanding summary of the research and theories related to regenerative biology at that time. Morgan's *Regeneration* appeared in 1901 and added to the active debates about how development works.[12]

Morgan saw the abilities of such organisms as earthworms, planarians, crabs, and hydra to regenerate after they were injured as providing a sort of natural experiment. For Morgan, these regenerative processes would provide a window on development, as historian Mary Sunderland has put it.[13] After he had presented the lectures, written several articles, and published the book, Morgan continued his exploration of regenerative biology. Because this work is now seen as providing a rather intriguing background to the current program on regenerative biology that dominates the National Institutes of Health, and because Morgan is best known for winning a Nobel Prize for his work on genetics and not for embryology, it is worth looking at his research in more detail.

Morgan spent nearly all his summers at the MBL as a researcher, a participant in the embryology course, and a trustee (from 1897 to 1945), and where his family went with him (Figure 3.4). Garland Allen, who has documented Morgan's work in the most detail, has discussed some of Morgan's contributions to development, but even Allen focuses on Morgan's genetics.[14] Morgan's developmental studies did not attract much attention until recently; now that developmental biology has gained more interest, the earlier work is worth examining, in part to see what ideas were left behind but might be useful. When we look more closely at Morgan's work a century ago from today's perspective, we find regeneration right at the center. We also find a particular vision for biological discovery that is informative about how those researchers regarded the role of science.

In particular, Morgan was fascinated by regenerative biology for two main reasons. First, the regenerative processes offered what he saw as an excellent means to understand development generally and in particular the phenomenon of differentiation. Second, regeneration showed a capacity to repair injuries that suggested a dynamic as well as a chemical and material response to changing conditions. This was surprising to some and intriguing to all. How might this dynamic capacity of repair work?

Morgan's *Regeneration* (which he dedicated to his mother) provided an excellent summary of the knowledge to date, starting with a historical chapter and running through the current experiments and theoretical interpretations. His broadly synthetic book, which brought together the study of many different organisms as well as a variety of methods, concluded that regeneration occurs through

Figure 3.4. Thomas Hunt Morgan with family and friends on the beach near the Marine Biological Laboratory. Morgan is standing in the center; his mother, dressed in black, is seated; Ross Harrison stands to the left and E. B. Wilson to the right. Other members of Morgan's family appear as well. See http://history.archives.mbl.edu for more on the MBL and Morgan. Courtesy of the Marine Biological Laboratory Archives; undated photo by Alfred Francis Huettner (b. 1884), "Thomas Hunt Morgan, E. B. Wilson, and others having a picnic." Available through http://hpsrepository.mbl.edu /handle/10776/2179. Courtesy of the Marine Biological Laboratory Archives.

internal factors that are useful to the organism and are part of growth and development.[15]

Researchers today have rediscovered Morgan's volume, after years of lack of interest.[16] Much less noticed and rarely read are Morgan's papers on regeneration, which show how he worked through his ideas before the book's publication and his continued explorations thereafter. Of the thirty-six articles he published that explicitly included regeneration in the title, from his first in 1897 to 1910, ten appeared in the MBL's *Biological Bulletin*, and another two appeared in the published set of MBL's *Biological Lectures*. That this was not merely a specialized interest of Morgan's alone is reflected by the full run of *Biological Bulletin* from 1887 through 2010 featuring 146 articles with "regeneration" in the title; for 856 reports, it is a keyword. This shows a remarkable continuing interest in the topic, and the research spans a diversity of organisms while addressing a range of questions and drawing on evolving types of methods.

Morgan's own studies of regeneration show that initially he explored a wide range of organisms. His essays describe research on hermit crabs, tadpoles, worms, ciliates, hydroids, salamanders, planarians, and other species (Figure 3.5). Morgan also led among American contributors in rejecting what they saw as simplistic Darwinian ideas about natural selection for explaining biological phenomena. In that view, variations arise, and some are more adapted to the environment than others; the more adapted are selected in the sense that they give rise to more offspring, and the traits that led to their selective advantage are inherited. Morgan saw this kind of interpretation as too theoretical and not sufficiently based on careful observation. Instead, a look at three of Morgan's papers reveals the range of his hypotheses and his careful empirical approach to science.

In "Regeneration and Liability to Injury," Morgan reported on research performed during the summer of 1897 in Woods Hole, studying the hermit crab *Eupagurus longicarpus*.[17] This particular case intrigued Morgan because of the attention it paid to the eminent

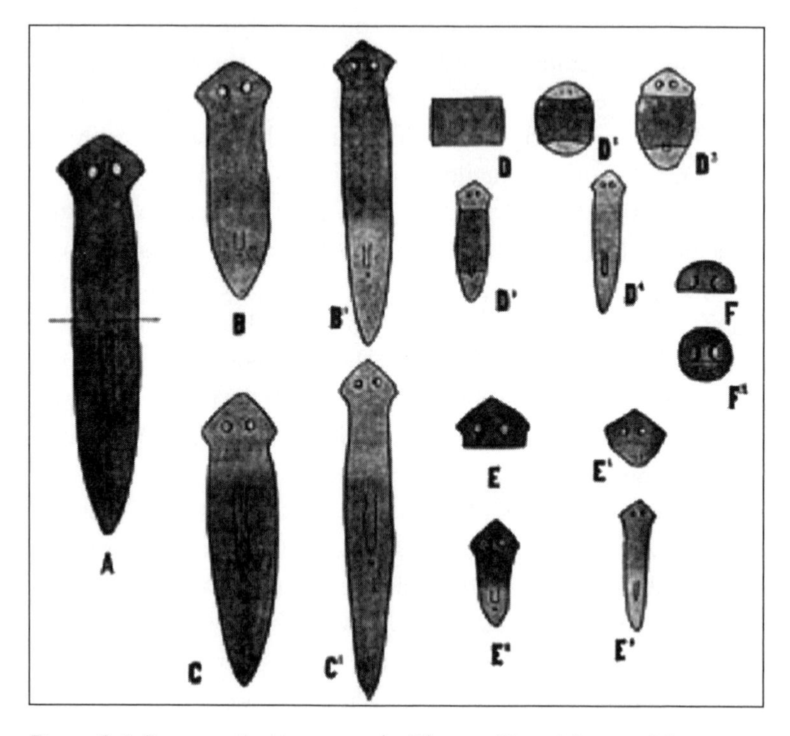

Figure 3.5. Regeneration in worms by Thomas Hunt Morgan. Morgan studied a wide range of species, cutting off parts and observing whether, when, and how regeneration occurred. Morgan's key to this figure: "*A–E: Planaria maculata. A:* Normal worm. *B, B1:* Regeneration of anterior half. *C, C1:* Regeneration of posterior half. *D:* Cross-piece of worm. *D1, D2, D3, D4:* Regeneration of same. *E:* Old head. *E1, E2, E3:* Regeneration of same. *F: P. lugubris.* Old head cut off just behind the eyes. *F1:* Regeneration of new head on posterior end of same." From Thomas Hunt Morgan, *Regeneration* (New York: Macmillan, 1901), Figure 4.

German researcher August Weismann's theories. Weismann had suggested that the parts most likely to be injured are those that have, through natural selection, developed a capacity to regenerate. Weismann's interpretation seemed plausible and tidy in some ways.

Yet Morgan denied this theoretical interpretation, which he felt went beyond and even contradicted what organisms actually do. Collecting crabs, he set out to find out what does happen. In particular, he started with empirical reports on which appendages were lost. This led him to ask, "Shall we find, then, in the regeneration of these different appendages any correspondence between the power of regeneration and the liability to injury or loss of a part?" Finding an answer required experimentation, in particular it involved cutting off different legs, antennae, eyes, and other parts. Some of the developmental process remained unclear: "In regard, however, to the problem of the frequency of injury of a part and its capacities to regenerate, the preceding results, I think, speak with sufficient clearness. No such relation is found to exist."[18] Those who postulated such a connection were misinterpreting the action of natural selection. Regenerative capacities do not require special circumstances but are part of normal developmental processes. It remained, Morgan suggested, to continue study of those processes.

In 1900, Morgan reported on experiments on tadpoles, in which he adopted the grafting techniques popular with MBL trustee and friend Ross Granville Harrison and Harrison's German counterpart Hans Spemann.[19] In work that might have been called tissue engineering, Morgan took pieces from the tails of one frog species and grafted them in different orientations to the tail of another very different looking species of frog so that it was easy to see which cells came from which species (*Rana palustris* and *Rana sylvatica,* to be specific). Morgan asked about the extent to which the regenerating part would develop out of the transplanted tissue rather than from changes in the host's own cells. Or could cells from both work together to produce new material and new processes? From many experimental results, he concluded that "a single tail may be formed by the regeneration of tissue derived from two species, and that in such cases there is no specific change produced in the one kind of new tissue by the other. Each kind of tissue regenerates its like, and the two kinds combine to form a single morphological organ—the tail."[20]

This work, which was used in his book on regeneration, carries two important ideas about development. First, regeneration is a natural process related to normal development and differentiation, not some special adaptation through natural selection just for particular organisms or particular parts. Second, cell fate is not predetermined by heredity or by the cell's position in the organism; rather, cells can reorganize with other cells to produce normal body parts in different ways. In *Regeneration,* Morgan laid out the experimental evidence and its interpretations, but he remained puzzled about *how* regeneration occurs.

In 1909, he turned to "The Dynamic Factor in Regeneration."[21] Looking at the hydroid *Tubularia,* he referred to his previous discussions of "formative factors" and acknowledged his earlier uncertainties. Was function necessary to produce regenerated form? he asked. Now he concluded, "That the dynamic factor in regeneration is not primarily the outcome of physiological movements of the animal, or of its parts, is made probable not only by many facts familiar to every student of regeneration, facts that show that the new part often develops under conditions where movement or function in this sense is absent. . . ."[22] Research on jellyfish by his students Charles Rupert Stockard and Charles Zeleny showed this as well. Several alternative chemical-material explanations might account for the phenomenon.

Morgan's paper shows his reasoning: he brings to bear evidence from different species that suggests that some sort of "formative stuff" plays a role in regeneration. Something internal to the organism responds to the changed conditions, including to injury, and helps guide a regenerative process. Morgan developed the idea of "morphollaxis" to explain the way that cells respond to injury and take on different roles and shapes to accommodate the new conditions. Of course, he realized that many questions remained about what directs the process in each experimental case, and about the effect of polarity and position in particular. Although he had previously been persuaded that both chemical-material factors and dynamic-physiological factors play roles, Morgan had by 1909 become convinced that some dynamic interactive physiological process was more probable for explaining the particular ways in which regeneration works.

His conclusion for the paper articulated a dynamic process for regeneration, stating that chemical changes can continue even when the stimulus that produced them ends. The changes just needed a stimulus to set the processes in motion, and then regeneration could continue. This was not a hereditary interpretation, as later researchers familiar with Morgan's work in genetics might expect, and he had already rejected the importance of evolutionary adaptation. Instead, Morgan was offering a very developmental, dynamic, internal, interactive interpretation. It was an interpretation that would be best pursued, explored, and tested by looking at a wide variety of organisms and performing a diversity of experiments. No limiting set of model organisms would do; rather, comparison and accumulating empirical evidence from across the animal kingdom would lead the way.

Though Morgan's focus moved for several decades to issues of genetics and evolution, he never lost his fascination with the fundamental phenomena of development. In 1934, his last book, *Embryology and Genetics,* brought him back to reflect on regeneration.[23] Morgan noted that because all cells have the same genes, "it would seem that every cell is capable of forming any part of the organism" except in cases where the tissue had differentiated so far that it could not adapt to new circumstances. "On this supposition it would seem more likely that the old cells would then continue to function in the new part as before, and, in fact, this is obviously true in many cases, but in the cells that change over into different tissues this explanation would not apply."[24] To Morgan, it seemed that the "location" of cells in the organism was the most important determining factor, and contact with existing tissue could provoke change in cells that were now in a different place or different circumstances than before.

Cells and tissues seemed driven to some extent and in some ways by hereditary factors or genes, yes, but even more by their interaction with other tissues. This interaction raised new questions for Morgan, and he concluded the last page of his final book by wondering about the role of genes in development. If cells could change in response to the environmental conditions of surrounding cells, then why could genes not change as well? That is, if the protoplasm of a cell "can change its differentiation in a new environment without losing its fundamental properties, why may not the genes also? This question is clearly beyond the range of present evidence, but as a possibility it need not be rejected. The answer, for or against such an assumption, will have to wait until evidence can be obtained from experimental investigation."[25]

Harrison and Spemann on Transplantation

Morgan and Loeb, along with Boveri and Wilson, worked on fundamental questions about life that had been raised by Roux and Driesch. They made the late nineteenth and early twentieth centuries a lively time for developmental biology, and all their basic questions remain important for developmental biology today. One other line of research, which added an additional perspective and complemented the studies of cells and regeneration, involved study of development using transplantation techniques. Researchers led by Spemann and Harrison asked, Why be limited to studying normal conditions with all the pieces in their usual places? Why not produce experimentally controlled conditions by moving the pieces around to see what they will do in different locations? Morgan did some transplantation experiments as well, but the others did far more.

Frogs are especially useful for "seeing" development through transplantation studies. They are plentiful, and they have many large eggs that are easy to collect and that continue to develop outside the mother so that all stages are visible. In addition, it is relatively easy to remove the jelly coat that normally surrounds the eggs and to snip off parts and move them to other locations during development. Furthermore, pieces from different individuals, or even different varieties or species can be removed and transplanted and still develop apparently quite normally.

Spemann asked what would happen if he removed pieces of the frog embryo—that is, if he simply cut pieces out and threw them away. Not that this experimentation was ever really simple because it required careful use of very sharp scissors and glass needles to perform the surgery just right. He asked, for example, how about removing the piece that is in the location that normally gives rise to the lens of the eye? This is called the area of the optical cup. Would the surrounding material compensate so that a lens still develops, or would the lens be missing? This got at some of the same sorts of questions that had drawn Roux and Driesch to their experiments: to what extent is differentiation self-determined by conditions within the embryo itself, and to what extent does it normally respond or can it under experimental conditions be provoked to respond to changing environments? Spemann discovered that when the piece is removed, the embryo is missing its lens and does not regenerate a new one. He continued with various other body parts, especially the buds that give rise to the arms and legs, and found that under many circumstances, the embryo does not adapt, but in some circumstances it does. This obviously raised many more questions about when, where, how, and why the embryo either compensated or did not.[26]

Spemann decided to try more radical interventions, which led to the concept of induction and of the organizer. He took pieces of tissue from one embryo and placed them deep inside the blastula stage (again, the normal stages had been mapped out already). Mostly nothing happened at all, and the host embryo looked as though it developed normally, even with that extra bit stuck in; the embryo apparently accommodated to the changed conditions. But in a few cases, the embryo adapted to the new material in an astonishing way—producing another, essentially whole embryo attached to the first. In work that Viktor Hamburger observed when he was a student in Spemann's laboratory, Spemann was keen to discover what made the difference between a resulting seemingly normal embryo and a resulting conjoined, duplicated embryo. What is it that allows some cases to develop a second new embryo in addition to the first? His laboratory sought to isolate the particular pieces that seemed to "induce" or "organize" this second embryo. In work performed with his student Hilde Proescholdt Mangold (and with some controversy about who contributed exactly what to the research, as Hamburger discusses in his reflections later), Spemann identified what he called the special "organizer" tissue, which he located as coming from the dorsal lip of the blastopore (Figure 3.6).

In the decades that followed, which Ross Harrison labeled the gold rush of experimental embryology, researchers tried all sorts of experiments to discover just what it was about the organizer that gave it the capacity to organize an additional embryo. Students and colleagues of Harrison and Spemann performed a wide variety of transplantation experiments with several frog species to determine what happens when pieces are removed or transferred from one place to another.

One widespread hope was that Spemann had discovered the material cause of differentiation. Ever since Roux's idea that chromosomal material is just divided up into different cells to bring about differentiation, it had seemed that the cause of differences among cells must be something related to causes within the organism itself rather than to actions from the external environment. Then when it became clear

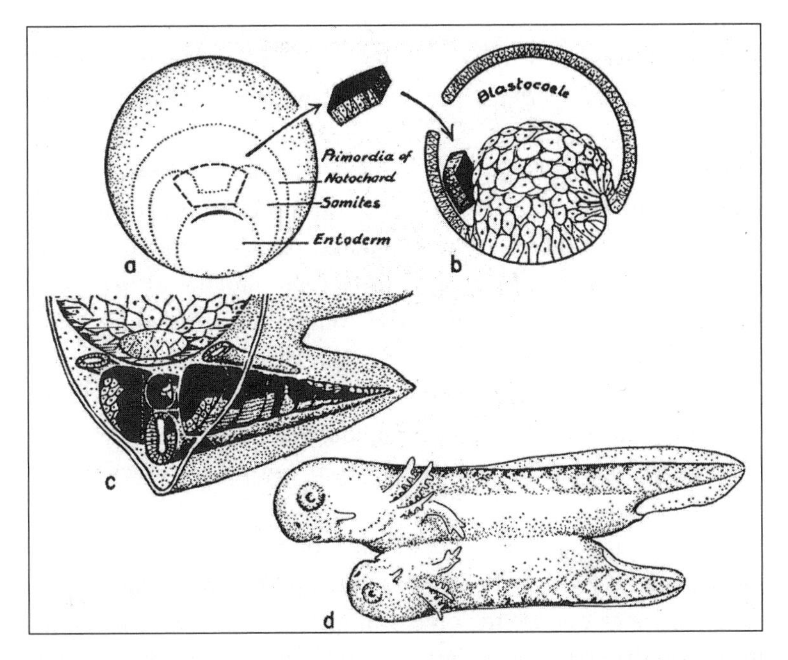

Figure 3.6. Image of Hans Spemann's organizer by Johannes Holtfreter and Viktor Hamburger, "Amphibians," in *Analysis of Development*, ed. Benjamin H. Willier, Paul A.Weiss, and Viktor Hamburger (Philadelphia: W. B. Saunders Co, 1955). In this egg from the frog species *Triturus cristatus* (a) the dotted outline shows the dorsal lip of the blastopore, which is transplanted inside another embryo (b) and gives rise to "organization" of a new embryo with chimeric somites as seen in (c) and to a newly induced embryo in (d).

that the chromosomes were not divided up into the different cells, researchers looked for other causes. Perhaps something internal, either a chemical property or a structural difference, could interact with the environment of the cells to bring about differentiation. The search was on to discover what the internal factor might be, with the idea that perhaps it was a very special set of cells that "induced" the development of the whole organism. That would be a lovely and easy explanation, it seemed, and perhaps Spemann had found it with that dorsal lip of the blastopore. But then Johannes Holtfreter demonstrated that Spemann's particular piece of tissue was not magical at all, and that in fact any number of inert substances could work just as well and could even induce differentiation of specialized neural tissue.[27] This realization, resisted at first, brought considerable disappointment to those who had hoped they were on to the secret of developing life, but Holtfreter had helped to keep the science of embryology firmly grounded in empirical materialism.

Harrison also performed a number of transplantation experiments, and his most surprising example involved transplanting tissue all the way outside the organism itself into an external culture medium—the first time anybody had done this successfully. He had a specific reason, namely, he wanted to answer a basic biological question that many others had asked: how is it that if development is gradual and epigenetic, the complex nervous system somehow figures out how to make all the connections work? Others had been arguing back and forth about whether the nervous system is actually a network that lies hidden but preformed inside the egg (somewhat like the materialists in the eighteenth century who insisted that the form must be there and we merely have not figured out how to see it yet). Harrison sided with those who held, in contrast, that the form arises gradually, and he thought of a way to test how it might happen. He transplanted neuroblast cells that under normal conditions would give rise to nerve fibers. He took them out of the frog embryos and placed them in a culture medium in a glass dish, using techniques borrowed from bacteriology to experiment with different culturing methods.

His experiment worked. In 1907 and more extensively in 1910, Harrison published the results demonstrating that nerve fibers can grow outside the embryo and do not have to be part of what some researchers assumed must be an integrated predetermined nervous system (Figure 3.7). Furthermore, they do grow in ways that seem remarkably like the patterns and processes they follow in normal developing embryos. This was the first ever successful tissue culture experiment in which the tissue lived *in vitro* (in a glass dish) rather than *in vivo* (in the living body). It was also the first stem cell experiment, as the neuroblasts he used are today known as neural stem cells, or in other words as those stem cells that are destined to become cells of the nervous system.[28]

Harrison's study of tissue culture, Spemann's of transplantation and the organizer, Morgan's on regeneration, Wilson's on cell lineage, Boveri's on chromosomes and the nucleus combined to yield a very dynamic and detailed picture of the developing embryo by the 1930s. Harrison continued with studies of regeneration and wound repair, including efforts to understand how tissues adapt to changing environments to regenerate function.

In 1938, Spemann suggested the possibility of a "fantastical experiment" to transplant a nucleus from one embryo into another cell from which the nucleus had been removed. And Morgan kept trying to discover how genetics informs development. This period of experimental embryology was one of tremendous discovery based on observations, manipulation, and a keen interest in explaining the phenomena of an unformed egg becoming a complex, formed adult organism through material, observable processes. The middle of the twentieth century shifted the focus toward inheritance and reoriented the way biologists and the public were understanding embryos, as we will discuss in more detail later.

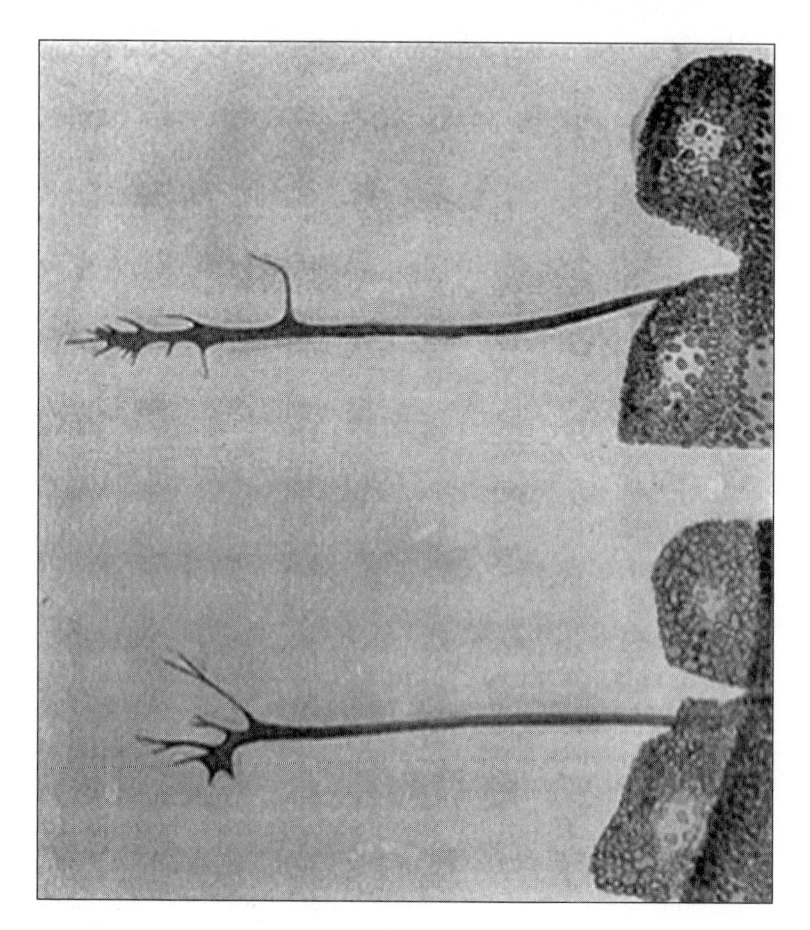

Figure 3.7. Ross Granville Harrison's illustration of nerve fiber outgrowth. Harrison used frog lymph as a medium in which to culture tissue, including a type of neural stem cell called neuroblast cells. He found that the cells grew nerve fibers in a way that resembled normal fibers. Ross Harrison, "Embryonic Transplantation and the Development of the Nervous System," in *The Harvey Lectures*, Series 4 (New York: Academic Press, 1908), 199–222.

Tissue Culture

Tissue culture provides a useful example of the intersection of science and its social impact. Ross Harrison performed the first successful culture of tissues outside the body when he transplanted neuroblast cells from frogs into an artificial culture medium. He used this technique to decide an embryological question about how nerve fibers develop and how development combines individual cells into a complex functioning nervous system.[29] First, Harrison determined that each nerve cell develops by a form of "protoplasmic outgrowth," in which the neuroblast cell reaches out into the surrounding medium and apparently uses that environment to direct its growth. Then, after he declared that he had achieved his own embryological research goals, he set aside tissue culture work and moved on to other embryological problems. But he had discovered an exciting new technology for biology and medicine, and Harrison encouraged others to take up tissue culture study. Of course they did, and Alexis Carrel was a leader in developing applications of the technique.

When the French surgeon Carrel came to the United States from France in 1904, he worked at the University of Chicago for two years then moved to the clinical environment of the Rockefeller Institute for Medical Research in New York. One of his colleagues there, Montrose Burrows, went to Yale to visit Harrison and learn about his new culture technique. At Yale, Burrows adapted Harrison's method of placing cells in a hanging drop, which Harrison had adapted from bacteriological methods, to cultivate tissues from chick embryos. When Burrows returned to Rockefeller to expand on his tissue culture exploration, he helped Carrel set up his own experiments with tissues from embryos and adults from a variety of organisms.

Where Harrison was driven to answer foundational biological questions, as a surgeon Carrel wanted to invent new treatments, and he particularly wanted to culture tissues for medical transplantations.

Carrel built on the techniques he learned to pioneer both tissue and organ transplantation in humans. In 1935, he and the aviator Charles Lindbergh (better known for being the first to fly across the Atlantic Ocean) invented a machine to allow organs outside the body to receive necessary fluids and support respiration. Their book *The Culture of Organs* discusses this invention and the clinical results it made possible.[30]

In addition, Carrel believed that he had cultured cells from a chicken heart that persisted over a long period of time (from 1912 to 1946, when his successors terminated the experiments), which he subsequently called "immortal." Carrel was so enchanted with the idea of immortal cells that he probably did believe that he had produced them. In retrospect, and even at the time, the fact that nobody else could replicate his results and the fact that it required such a long leap from observed culturing for several generations to the supposed immortal cultures should have raised questions. The most likely explanation is that every time he or his assistants changed culture mediums, they unknowingly included some fresh new cells, though some critics alleged that the assistants added new nutrients on purpose. Debate continues about how to interpret this apparent cell immortality.

What really stands out in this case is the passion that Carrel and others had for the result and its medical implications.[31] They were eager for the clinical advances that long-term cell and tissue culture would make possible, and they even foresaw the culturing of organs for transplantation. Many decades later, we know that some kinds of cells really can become effectively immortal, in the sense that they can be cultured indefinitely. This has occurred with some abnormal cell types, such as the HeLa cancer cell from Henrietta Lacks, which Rebecca Skloot discusses in her astonishing book *The Immortal Life of Henrietta Lacks*.[32] It also happens with embryonic stem cells, which are defined as self-renewing and also as immortal under some circumstances.

For the background to this idea, in addition to Carrel's efforts to culture cells and tissues in a clinical setting, we can look back at Ross Harrison's final published research of his long and distinguished career. In 1938, Ross Harrison retired from his professorship at Yale and served as chairman for the National Research Council from 1938 until 1946, during the critical war years. He also joined the Science Committee of the National Resources Planning Board, and chaired President Roosevelt's Committee on Civil Service Improvement in 1939. After the war, he returned to research with his last new paper on wound healing.[33]

In this article, Harrison returned to issues of regeneration, looking at frog embryonic development and asking about the experimental conditions under which the embryo recovers its structure and presumably its function after injury. The central nervous system, Harrison began, seemed to have "considerable powers" to heal and reconstitute the normal, even in the face of experimental intervention and damage.[34] He focused on the development of neurula-stage embryos (when formation of nerves begins to take place), and he examined the repair of wounds at that stage. What happens when an injury occurs? As tissue moves to cover the wound, does it come from completely new tissue generated at the wounded edge? Or does existing tissue migrate from another location to cover the site? Is the recovery complete, or does it incorporate adaptations to the new conditions?

Harrison asked what happened in this abnormal pathological case of tissue injury, and he also came to the conclusion that a better understanding of normal conditions would help him understand how wound repair works under abnormal conditions. In addition, he realized that understanding wound repair would also inform our understanding of normal developmental processes. Fortunately, as Harrison had already plotted the standard sequence of normal frog developmental stages back in 1900 (with copies of his plates distributed to every student), he was well prepared to make comparisons against the newly created abnormal wound conditions.

One of the most active researchers in the field of tissue repair and development was Harrison's former doctoral student Samuel Detwiler. Detwiler, who received his Ph.D. in 1918 and named his second son after Harrison, followed Harrison and Spemann in his own research interests, with a focus on neural development.[35] Detwiler asked questions about how wounds are repaired through regeneration of tissues, and Harrison took up those questions in the context of his own questions about regeneration, which he had pursued earlier in his tissue culture studies. As with his earlier studies of nerve fiber outgrowth, Harrison concluded again that what happens in the experimental case is very nearly like what happens in normal development.

Harrison thought in terms of structure and function of the living embryo and its parts, and not in terms of genetic regulation as many researchers would at least include in the discussion today. Presently, regeneration researchers look for such genetic factors as up and down regulation, regulatory response, knockout, knockdown, RNA interference, markers, and indicators of signaling pathways along with other tools to help discover and interpret results of underlying genetic causes of regenerative capacity. Not only did Harrison have none of these tools available, he also did not view the organism in those hereditary terms. To him, the structured tissue moved as cells migrated, and he wanted to map out what happened morphologically to the form, step by step, and to interpret the developmental changes in physical and mechanical terms. His approach helped advance our understanding of embryos as well as providing grounds for imagining how we might reshape developing embryos in the future. Others have picked up his structural emphasis as well.

Differentiation

One persistent question through all of these lines of experimentation concerns what causes differentiation, particularly the relative importance of direction by the nucleus or cytoplasm and the relative roles of genes or environmental conditions. We have already seen how this discussion played out at the end of the nineteenth century.

Another central developmental question for the twentieth century concerned the cause of growth, which led to the questions what causes cells to grow larger and also what causes them to differentiate, or to grow *as* some particular kind of cell? What causes growing cells to become differentiated as nerves, for example; what helps direct their paths as the neural cells stretch out into their surroundings? An important part of the story depends on something sensibly named nerve growth factor, which plays a central role in promoting growth as nerve cells. Discovery of the nerve growth factor brought a Nobel Prize to some of its discoverers. In addition, the identification of this particular growth factor led to the discovery of a rich diversity of other growth factors. The eventual recognition that these factors play an essential role in differentiating cells into different kinds of tissues and allowing organisms to become complex systems of interacting parts added an important new dimension to understanding how embryos develop.

Viktor Hamburger's laboratory at Washington University in St. Louis, Missouri, provided the nurturing environment for work on nerve growth factor (NGF). Educated in Germany under Hans Spemann, Hamburger came to the United States along with many others escaping the Nazis. He went first to the University of Chicago with the help of a Rockefeller Fellowship in 1932 and worked there with Frank Lillie. In 1935, he moved to St. Louis, where he remained for the rest of his long life until he died in 2001 shortly before his 101st birthday, still publishing until the end. At Washington University, Hamburger became a leader, including service as chair of the zoology department (later called the biology department) from 1941 to 1966. In that role, he successfully persuaded the university to expand the department so that it became outstanding in biological research.

As Hamburger explained in the abstract to an article about the history of the growth factor discovery, NGF was the first of a sequence of other growth factors that followed. He acknowledged (and it is worth seeing this episode in his own words) that:

> The history of its discovery is very colorful; it is a rare combination of scientific reasoning, intuition, fortuities, and good luck. In addition, I believe that the collaboration of three scientists with very different backgrounds contributed to the success: I had grown up in a laboratory of experimental embryology, Dr. Levi-Montalcini came from neurology, and Dr. Stanley Cohen was from biochemistry. The decision where to begin the history of a discovery is always arbitrary. I shall give my reasons why I begin this story with my wing bud extirpations on chick embryos and the analysis of the effects of the operation on the development of spinal nerve centers, published in 1934. Of course, I am aware of the fact that the analysis of neurogenesis had been pioneered by Dr. R. G. Harrison and his students at Yale University since the beginning of this century. It should be mentioned that their experiments had been done on amphibian embryos. My own interest in problems of neurogenesis dates back to my Ph.D. thesis in the Zoology Department of Professor H. Spemann at the University of Freiburg in (the Federal Republic of) Germany; it dealt with the influence of the nervous system on the development of limbs in frog embryos. After I had obtained some inconclusive results I did the crucial experiment of producing nerveless legs. I removed the lumbar part of the spinal cord and the spinal ganglia before the outgrowth of nerve fibers. The nerveless legs developed normally in every respect, but the muscles atrophied eventually.[36]

This account shows Hamburger's perspective on a discovery about which much disagreement arose about exactly who did what leading to the NGF discovery. In fact, only Cohen and Levi-Montalcini received the Nobel Prize, and many feel that the prize committee had slighted Hamburger in not also naming him. The Nobel committee can name no more than three, but for this 1986 award they named only the two. The selection led to much controversy, though as Hamburger's letters show, he tried very hard to stay above the battle and away from accusations.

Levi-Montalcini, who died while still active at age 103, was the oldest living and the longest-lived Nobelist at the time of her death. At times, especially early on, she generously acknowledged the contributions of others to her research. At other times, and more recently, she seems to have enjoyed telling the stories about her wartime hardships and her struggle as a woman and a Jewish scientist to perform research at all. She tells of collecting eggs from neighborhood farms and studying them in her bedroom, inspired by Hamburger's reports of cutting limb buds off chicks to determine what would happen. She presents herself as a lone heroine drawn by the excitement about science and exhibiting great originality. No doubt, all the stories are partly accurate. Yet science works through collaboration and teamwork, and she clearly gained tremendous insights and expertise in Hamburger's laboratory.

Eventually, after the war, Levi-Montalicini made her way to St. Louis at Hamburger's invitation. The laboratory continued studies of what conditions promote growth and differentiation under normal circumstances, but also added examination of what happens under experimental conditions. The earlier studies on cells had shown much about the way cells grow and differentiate but little about what mechanisms regulate the growth of cells and organs. This group discovered that, in fact, a particular and identifiable factor promotes growth. The effect became clear when Levi-Montalcini transplanted cells from mouse tumors to chick embryos. The mouse tumor cells caused the chick nervous system to grow quickly and dramatically, which raised the question of why. Further experiments showed that even a small amount of NGF would induce rapid cell growth. By 1952, Levi-Montalcini had made clear that there was some growth-promoting material factor.

After the biochemist Stanley Cohen joined the group in 1953, they purified the apparent growth factor and began to analyze the amino acids that make up the protein involved. In later studies, researchers discovered that NGF actually plays additional roles beyond promoting growth; it contributes to the survival of some neurons and serves as a signaling molecule for others. The work in Hamburger's laboratory stimulated the discovery of many other growth factors and related molecules that do much of the work of regulating development. Any attempt to engineer organisms absolutely depends on understanding how these factors work; research on growth factors gets at the mechanisms involved, not just the theoretical ideas or inferred relationships involved in Spemann's rather vague idea of embryonic induction. This research represented a major advance toward understanding and then controlling the material processes of development and differentiation.

Applying Science to Improve Lives

We have seen that biological researchers led by Spemann and Harrison had developed techniques for transplanting bits of embryos, including the parts that normally give rise to limbs, eyes, ears, or other pieces, and their research illustrated what occurred when researchers carried out particular transplantations. Harrison went even farther, transplanting cells out of the organism altogether. Carrel worked on transplanting organs, and others tried transplantation of various tissues for medical reasons.

In the 1950s, the transplantation of hematopoietic stem cells introduced a new approach and eventually helped lead to new discoveries about how development works. Hematopoietic cells are a special kind of cell found in the bone marrow that can differentiate into several different kinds of cells, including blood cells. This makes these cells useful for treating blood disorders such as leukemia. In that postwar decade, researchers experimented with transplanting bone marrow into patients who had been irradiated, had leukemia or other blood diseases, or had extreme burns or other injuries that involved severe disruption of blood production and flow. Through trial and error, as is often the way with clinical applications, it became clear that these particular bone marrow cells have the capacity under some conditions to give rise to red blood cells. But the transplants only rarely worked, most commonly from one identical twin to the other. This suggested a rejection response by the recipient, which limited the usefulness of the technique.

Fortunately, the late 1950s brought the discovery of the proteins that allow the body to determine whether cells are one's own or from a foreign source. With that information came the first steps toward developing antirejection drugs to allow the body to accept foreign cells. These treatments involve complex, expensive, and often demanding management, but they make tissue and organ transplantation possible when it could not occur otherwise. In 1967, the surgeon Christian Barnard transplanted a heart into a patient named Louis Washkansky in South Africa. Even though Washkansky died eighteen days after the surgery of pneumonia, his heroic willingness to endure the experiment established a beginning for organ transplantation. By 1973, the first successful bone marrow donation took place from one donor to an unrelated patient. Other tissue and organ transplantations followed, with increasing success. Yet even after Congress passed the National Organ Transplant Act in 1984, patients in the United States and elsewhere have faced a serious shortage of available organs and tissues. Thousands of patients die every year for lack of appropriate donations. Engineering tissues to increase acceptance or constructing parts to replace function offer both exciting prospects to provide alternatives and a source of additional transplantable tissues and perhaps even organs.

One assumption about the capacities of transplanting parts has been that an organism operates as a defined individual and that once it gains its individuality as a particular kind of thing, it stays that way. It might be possible to move around the parts of lower organisms such as hydra, or even to take apart, separate, and reaggregate the cells of sponges (as Henry van Peters Wilson at the University of North Carolina had shown in 1907 and many have observed since). But surely this putting together of parts cannot work in more complex organisms, and especially not in mammals. Yes, transplantation of particular organs or tissues might be possible to replace lost function, but biologists found it hard to imagine success with anything more radical. Yet this hard to imagine possibility had already begun to become reality when researchers showed first with mice and then eventually with humans that they could transplant cells. We will return to these themes of engineering and constructing embryos in later chapters and will look to the future as well.

4

Inherited, Evolved, and Computed Embryos

So far, we have focused on the history of people looking at embryos and describing what they saw. They have experimented with embryos, developed ways to look inside them using microscopes and other tools and techniques to see more detailed structures, and then have tried to make sense of what they have seen. Yet the focus has remained largely on the physical object, with discussions about causation based on understanding how one stage gives rise to the next through material and largely structural or sometimes physiological changes.

Many of the descriptive and experimental embryologists we have discussed at the turn of the twentieth century and well into the twenty-first century were not concerned with heredity or evolution. They were focused on the developing embryos that they saw through the microscope in front of them and were not concerned with a background of more distant evolutionary causes that might contribute to directing development. Though they obviously understood that heredity occurs, as offspring inherit some features from the egg and sperm cells of parents, they did not have clear and widely accepted theories about how that occurs. And though they universally accepted that evolution occurs, they either did not believe that it directly affects development or did not see any effective means of studying how evolution affects development. Their focus was on the embryos themselves.

Meanwhile, the study of genetics was progressing in parallel, largely ignored by most of those who called themselves embryologists. It was not until the 1950s and 1960s that an increasing number of embryologists embraced the science of genetics and began to investigate the underlying genetic information that might be directing development. They began to ask how genetic factors, which are presumably located on the chromosomes, might influence how embryos respond to their changing environmental conditions. In the 1960s also, visionary theoreticians began to imagine how even more distant evolutionary history might have shaped heredity and combined with developmental constraints to direct epigenetic development. These different fields of research—kept largely apart because of their different methods and questions as well as their separate working locations—began to come together in promising ways.

Inherited Embryos

Most educated people learn in school that James Watson and Francis Crick discovered that DNA has the structure of a double helix, and that they received a Nobel Prize for their work. When they announced this discovery in 1953, they said, briefly and pointedly, that "It has not escaped our attention that the specific pairing we have postulated immediately suggests a possible copying mechanism for the genetic material."[1] Theodor Boveri had already demonstrated the importance of the chromosomes for development, and others had later showed that the chromosomes consist of DNA. Morgan's laboratory group had demonstrated connections between chromosomes and genes. Thus, development connected to heredity and DNA in a concrete way.

Further, Watson and Crick showed that the chromosomes are made up of DNA, with two strands of the molecule wound around each other, and that they can unwind to allow reproduction. This allows each of the simple strands to serve as a template to make another matching single strand or a new double version. In 1953, Watson and Crick and their DNA model appeared on the front cover of magazines and newspapers, and DNA became a biological darling. These days, DNA seems to appear everywhere in public: in advertisements for perfume and cars, in the medical literature, in phrases such as "sports (or music or whatever) is in our DNA," and even in offers to have your own DNA mapped through companies like 23andme.com.[2] The chromosomal strands of DNA are understood to carry the genes that code for physical traits, once the developmental process has caused "expression" of the inherited information.

Not surprisingly, with increasing emphasis on chromosomes and DNA, perceptions shifted away from seeing embryos as starting from clusters of dividing cells without form and driven by epigenetic responses of internal structure and to looking at external environmental conditions. The focus shifted more toward heredity and predeterminism, in which developmental processes are seen as defined and directed by inherited information in the DNA. Cells continued to be a major focus for studying development because they carry the inherited chromosomes and also undergo division that leads to differentiation and growth. Thus, cytology combined with embryology in the study of development.

In the late 1950s and 1960s, embryology became widely known as developmental biology, with a focus on genetics and development and their interactions. Professional organizations emerged and adapted; the Society for the Study of Development and Growth Symposium grew dramatically in size and became the Society for Developmental Biology. Embryology as a field continued, of course, to look at the structure and function of embryos, but it resided largely in medical schools where the focus remained on health or on problems of embryos for medical reasons. Fewer and fewer biology programs offered courses in embryology; even those that remained would typically cover an increasingly wider range of developmental topics, including developmental genetics.

This enthusiasm for genetics was not universal. For example, one of the leading embryologists who had played a dominant role in the field and who lived into the 1950s did not see the point. Ross Harrison, then retired from Yale, knew that heredity obviously happens, but he did not see that embryology had the tools available to make sense of heredity's role. By the 1910s, Harrison had been one of the most important biologists in the United States and directed the biology department at Yale University. There, he built a grand new laboratory, revised the curriculum for undergraduate and graduate students, trained a large number of very successful graduate students, and almost surely would have won a Nobel Prize had World War I not intervened just as his nomination went forward. Harrison played an important role as the president of various professional societies, the editor of the *Journal of Experimental Zoology*, and the head of the National Research Council. He remained a scientific leader throughout the 1940s.

Yet even though he had access to the best biological knowledge of the day, Harrison did not think genetics mattered for understanding development. The embryo starts as a fertilized egg, and what mattered was how it develops from there. What constituted development, according to Harrison and his like-minded colleagues, was its morphological structure and the way it gradually emerged and changed through morphogenesis and the traditional embryological processes of growth and differentiation.

As department chair at Yale, Harrison repeatedly rejected the suggestion by younger colleagues that the department should hire a geneticist. Not only did he not see genetics as important for embryology, he did not see that field as making much progress in addressing central biological problems of any sort. Even his good friend since graduate school at Johns Hopkins, Morgan, could not persuade Harrison that genetics mattered enough to hold a major place in biological programs.[3] Harrison saw room for molecular biology, biochemistry, evolution, and ecology, but not for genetics.

A graduate student at Yale, Jane Marion Oppenheimer worked directly with Harrison's student John Spangler Nicholas and also indirectly with Harrison. She trained as an embryologist, called herself a developmental biologist, became an excellent historian of embryology, and talked of how the field of developmental biology was shifting away from the experimental embryological approaches of the distinguished but retired Harrison.[4] Starting in the late 1950s, researchers who had emphasized phenomena of growth and differentiation began to ask about underlying causes. As scientists, they had always asked why the cells and tissues develop as they do, but the answers had typically focused on what about the immediately preceding stage had caused the change to the current stage. Now they could ask a deeper why question: why does the process develop in this particular way? What is it about the chromosomes and the genotype (which is made up of all the genes) that causes the information there to be expressed in this particular phenotypic pattern (or the physical traits that we see)? How does the genetic potential become actualized as the phenotypic organized organism? The graduate students and younger researchers saw the importance of genetics, even if the older generation of leaders like Harrison did not.

The new emphasis on genetics led many researchers to focus on the chromosomes and the nucleus, so all that careful study of cell divisions that Wilson and Boveri had carried out a half century earlier now gained new importance. Exactly how do cells divide, and exactly how does the inherited material duplicate so that every cell has the same set of genetic material? And because every cell has the same chromosomal inherited material, as Lillie's paradox had recognized earlier, what causes the cells to become different from each other through their differentiation? What mechanisms, in other words, drive cell division? And what processes of differentiation bring development?

Most researchers now looked at the nucleus and genetics in terms of chromosomes, but some realized that hereditary information also resides in the cytoplasm as well as in mitochondria. However, even among those enthusiastic about studying genetics, not everybody rushed to embrace the nucleus as directing development. Indeed, important biologists continued to study cytoplasmic inheritance. For example, Tracy Sonneborn spent much of his career at Indiana University studying the protozoon known as *Paramecium*. With these relatively simple organisms, Sonneborn demonstrated the impact of non-mendelian and non-nuclear or cytoplasmic inheritance.[5] Where Mendelian inheritance emphasized the regular, predictable probabilities of an offspring inheriting a particular allele, which presumably then translated into expression of the particular dominant gene, non-mendelian inheritance did not follow this pattern but rather showed considerably greater variability. For most researchers, this additional way of seeing heredity did not play a major role, but it is important to know that researchers were also exploring other ways heredity could occur and what such non-mendelian or cytoplasmic inheritance might mean.

Developmental Genetics

For the study of development, the stories start "in the beginning there was Aristotle," but for genetics it is "in the beginning was Mendel." The work of the Austrian monk Gregor Mendel (1822–1884) overlapped with the height of Darwin's research, and biologists often lament that Darwin did not appear to have paid any attention to Mendel's ideas, even though he may have seen a copy of Mendel's paper on pea breeding. In retrospect, historians have pointed out that even if Darwin had known of Mendel's research, he would likely not have known quite what to do with it. Darwin's view of organisms and embryology emphasized continuities and wide-ranging variations rather than fixed, discrete, discontinuous differences.

Mendel worked in an Augustinian monastery where he performed breeding experiments in plants. According to historical records about Mendel (sadly many of which were destroyed by a misguided successor at the St. Thomas's Abbey in Brno, Czechoslovakia, where Mendel worked), he initially had been interested in breeding in mice but was persuaded to study plants. The Abbott of St. Thomas not only encouraged Mendel to continue his studies at the University of Vienna, but he also set aside land for Mendel to cultivate. Mendel chose peas, specifically *Pisum sativum,* and he crossed different varieties by carefully controlling their pollination. He carried pollen from one plant to others by hand, over and over again, to study the results in a population of peas. He maintained that there would be regular, predictable patterns in the traits expressed. For example, some individual peas were wrinkled and others smooth, while some were lighter color and yellow and others were green, and so on. He picked out seven traits and studied them meticulously. His traits all came in distinct and discontinuous pairs—the peas were either wrinkled or smooth, yellow or green, and never some combination of the two varieties.

Mendel found that the results were predictable, but this came as a surprise to many breeders at the time. For example, when he crossed peas that were always wrinkled with those that were always smooth, he found that the peas of the next generation were all smooth. When he made his crosses a second time, the experiment yielded a second generation of offspring in which one-quarter were wrinkled and the rest smooth. They were not in-between, but were either one or the other—wrinkled or smooth—which meant the characteristics of wrinkled and smooth did not blend but remained segregated. They appeared in this three-quarters to one-quarter ratio quite reliably, thereby showing that whatever was causing the particular trait remained constant. Mendel called these causes "factors," and they correspond to what would later be called "alleles," which are considered different forms of what would later be called "genes," which provide information for each trait. Mendel concluded that one dominant factor led to smoothness, one recessive factor led to wrinkledness, and when one of each factor came together in the hybridization process between two varieties, only one would prevail over the other.

Then in the next generation, it is still the case that only one factor prevails, but the recessive factors for wrinkledness have not gone away altogether. The factors were just that: recessive. They persist in the cell and, in fact, even re-exert their effect when two recessive factors come together. But because they are recessive and it is the dominant factors that are expressed, these recessive traits and their factors are invisible in the first generation. Because each of the crossed plants actually has both a factor for wrinkled and a factor for smooth, Mendel could predict that the two wrinkled factors would come together in precisely one-quarter of the offspring. Another one-quarter would have two smooth factors, and would be smooth. And one-half of the offspring would have one of each factor, or be hybrid themselves as the first generation of hybrids were, and they express the trait of the dominant factor.

Furthermore, Mendel concluded by studying a particular set of seven different characters that each sorts independently of the others. That is, whether a particular factor leads to smoothness does not affect whether the plant also has a factor that leads to yellowness or whatever. It turns out that he had selected, clearly not by accident but clearly also not with any knowledge of the chromosomes involved, seven traits that are controlled by different factors that are very far distant along the same chromosomes in a few cases, or physically on different chromosomes altogether in other cases, and therefore not linked in any way. Mendel concluded from the regularities he saw in the characters of his pea plants that his results could be characterized as laws: the law of segregation and the law of independent assortment.

According to some historians, Mendel was a revolutionary who did not cause a revolution. His experiments were beautifully conceived, meticulously carried out, brilliantly interpreted, and largely ignored. Mendel was not an established researcher, he had no higher degrees, and he was suggesting something that did not fit the standard views of the time. The other plant breeders of Mendel's time assumed that the hereditary material from each parent could blend together. Or if one parental influence prevailed, there was no reason to believe that inherited material from the other parent remained intact. The idea of distinct and defined underlying inherited factors that determine the characteristics that we see was a new idea.

The story about how Mendel's work was rediscovered has been told well and often. What is important here is that when his work was rediscovered in 1900, it took hold. Mendelian genetics did not become standard immediately, but the idea of inherited discrete units certainly had to be considered. One of Mendel's rediscoverers, William Bateson, coined the term *genetics* to describe this study of heredity. Then, in 1909, Wilhelm Johannsen clearly distinguished the inherited gene and the genotype, which included the collection of all the individual's genes, from the phenotype, or the characters that we actually see, which are the expression of the genotype. Johannsen established the language used thereafter, and Mendel's factors largely dropped into the background.

In 1910, Thomas Hunt Morgan was continuing his studies of development and had become intrigued by this discussion of heredity and development, as he was one of the relatively few closely studying evolution, embryology, and heredity. Just how are these fundamental biological processes all related? he and his contemporaries were asking. What is the role of Mendelian genetics in determining or guiding development? In one paper published in 1910, Morgan rejected the Mendelian ideas. He did not like the idea of hypothetical genes; Mendel had not actually seen any such thing as factors, and certainly he had not seen them segregate as his interpretation suggested they must do. In his article in *American Naturalist,* Morgan therefore rejected the idea as mere theory. But as one to keep an open mind, Morgan saw that some kinds of discoveries could count as evidence for the Mendelian theory.[6]

In fact, also in 1910, the evidence appeared—and he found it in his own laboratory. Morgan was an opportunist in the very best sense of that word. As we have seen with his studies of regeneration, he would use whatever organism seemed promising. In 1910, he was experimenting with the fruit fly *Drosophila* that colleagues had suggested he should study as a way of examining evolution. He and his laboratory group looked at a lot of these little flies and noted the variations they found. The males always seemed to have red eyes, so they concluded that that was the normal type. Then they found a white-eyed male. Many observers might have set that one aside, but Morgan saw it as potentially important; he crossbred it with red-eyed, normal sisters. He published the results in a special short paper (the type of rare publication that appears occasionally in scientific work and is read forever after and referred to as a classic).

Morgan reported that the breeding experiments with the white-eyed male and the red-eyed females yielded 1,237 offspring. All of these had red eyes except for three with white eyes, which were all male. In reflecting on the results, he concluded that the red eyes were the dominant result, with presumably a constant dominant cause underlying the trait. He thought that the very rare additional white-eyed males must come from "sporting," what we now call mutations. Inspired by Mendelian theory, he then crossed this generation to see what would happen in the second generation. He got 2,459 red-eyed females, 1,011 red-eyed males, 782 white-eyed males, and zero white-eyed females.

His results did not fit exactly with a Mendelian interpretation. Why were all the white-eyed flies male? Morgan hypothesized that the gene for white eyes was not entirely independent of all other factors in heredity. Rather, it appeared to be sex limited. By 1910, research had shown that sex is determined by chromosomes, so if white eyes are connected with sex, and sex is connected with chromosomes, then white eyes (and, by inference, other inherited factors for other traits) are connected with chromosomes. In other words, chromosomes are the material carrying the information of heredity. His laboratory group continued to explore these connections, and performed a comprehensive mapping of the chromosomes. By showing how often traits were linked together, they could infer how closely spaced the genetic factors were along the chromosomes. In 1933, Morgan received the Nobel Prize in Physiology or Medicine for this work linking chromosomes, genes, and traits of the organism.

Heredity

By the time that Watson and Crick showed the structure of DNA, in the context of understanding that chromosomes consist of DNA, the pieces that make up the puzzle of how heredity and development work together were beginning to come together. Traits that develop in an organism start out with inheritance of genes that come on chromosomes, and these interact with environmental and other as yet unknown conditions. But this would not have happened without a further understanding of the role of chromosomes.

Oscar and Richard Hertwig, Boveri, Wilson, and others all contributed to our understanding of the complex processes by which the cell divides. Through the process of *mitosis,* a cell can divide into two daughter cells that have the same number of chromosomes, though some of the chromosomes come from the mother and some from the father. In the case of germ cells, the cells divide through a process of *meiosis,* in which the resulting eggs and sperm will have the right number of chromosomes to allow them to combine during fertilization. The work of these researchers showed that the chromosomes were necessary for reproduction as well as for normal development. But it was not clear precisely what the chromosomes actually do. Studies of sex determination had already provided the first major hints of concrete action, though the causes remained unclear.

Into the early twentieth century, the public assumed that the mother's behavior largely determined the sex of her offspring. Biologists wanted a more concrete and material explanation, and their hypotheses largely focused on environmental conditions as shaping the results. But others were looking instead to internal factors as most important, and Mendelian genetics suggested that some kind of gene might play a role. Because sex is much more complex than just one trait, just one gene or factor did not seem likely to do the job. Walter Sutton at Columbia University and Boveri suggested that the chromosomes themselves might determine sex. In some species, the males have an extra chromosome; in some, the females have an extra; and in others, there is a chromosome that looks very different in males and females. Obviously, this work required examining a lot of chromosomes, which in turn required more effective techniques. We have discussed Wilson's careful study of cells, and it is not coincidental that Sutton's work was done in Wilson's laboratory, as was the work of Nettie Stevens and other important contributors to the Mendelian, chromosomal theory of heredity of sex.

Something about the chromosomes determines which sex an individual will become. Chromosomes contain the genes that are the material units of heredity, and heredity determines development. It has taken a very long time and will continue to require more research to understand the mechanisms by which this translation of inherited genes on chromosomes into expressed traits takes place, which will bring us right back to debates about epigenesis and preformation.

Some researchers began to look for evidence of the action of genes and to map genes on chromosomes. Many hoped that they could find the "gene for" or "chromosome for" whatever condition they were concerned about. It is worth noting that not everything was about genetics, however. In 1916, Frank Lillie published his work on the freemartin and showed that production of sex characteristics is not all determined by genes. The freemartin, which occurs among cattle, is a female that is very masculine in features and is sterile reproductively. Breeders wanted to know what caused this condition and how to avoid it. Already some were speculating about genetic causes, while others pointed to environmental conditions—which seemed controllable. Lillie hypothesized that the freemartin is a female in the presence of a fraternal male twin (Figure 4.1). He reasoned that the two each had a chorion (the membrane that separates the mother and fetus) connecting the fetus to the mother to receive nutrition, and that in some number of cases the chorions could become twisted and connected in a way that would allow fluids to move from one to the other. The female was already chromosomally on its way to becoming female, but hormones from the male essentially took over the female's process of developing sex characteristics. Lillie noted that the evidence was insufficient to determine whether the male sex hormones are more dominant, or whether they just acted sooner and thereby took over the female's system.[7] He pointed to other examples where such twinning led to feminization rather than masculinization.

Such work as Lillie's showed clearly that both inheritance and environmental conditions, in this case the intertwining of hormones through the chorion, play important roles in shaping development. Where "nature" was seen as representing heredity, and "nurture" meant response to the environment, both were seen as working together, although one might dominate. The nature of the interactions

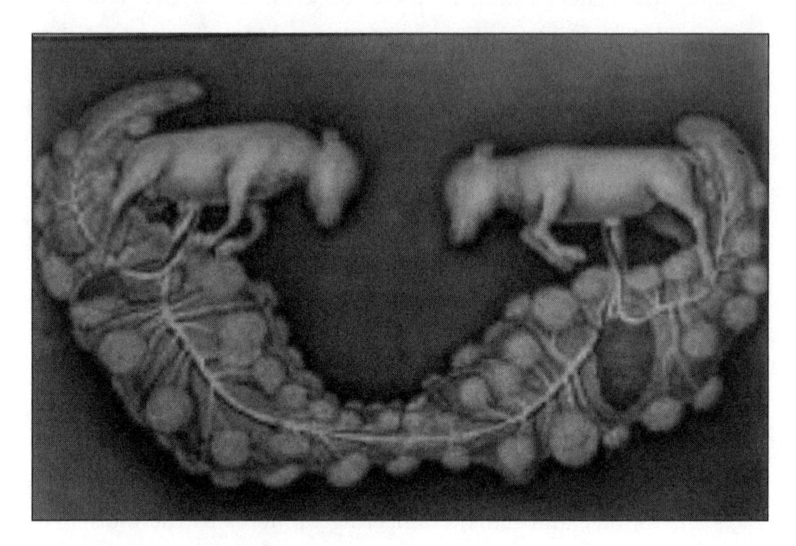

Figure 4.1. Frank Rattray Lillie's freemartin twin model. The chorion membrane is connected such that fluids move from one twin to the other; when the male twin's hormones overwhelm the female twin's development of sex characteristics, a sterile masculinized freemartin is produced. The Marine Biological Laboratory has a print of this image identified as "Original painting of Free-martin twins by K. Toda (1916). See Figure 4 (p. 426) Lillie, *Jour. Exp. Zool.* 23 (1917). Given to B. Lt. Willier in March 1937, 'to keep for posterity.' Given to James D. Ebert in 1971." This gift was for the laboratory's Rare Books Room. Used with permission of the Marine Biological Laboratory, Rare Books Room and Archives. My thanks to Diane Rielinger.

of the two as well as the causes and processes that affect the balance provided fodder for much debate. This also led to extremists on both sides: those who argued that everything about development is caused by reactions of individual organisms to the environment, and those who focused only on inheritance and the information of the genes. The debate continues today, as Evelyn Fox Keller discusses in her work *The Mirage of a Space between Nature and Nurture.*[8]

Although the previous chapter focused on the morphological studies of epigenetic embryologists, here we return to those areas of research in which predeterminism prevails. If the genes actually do cause the ensuing developmental results, then the developmental processes are largely directed by the genes. But how?

DNA and Determinism

Recall what Watson and Crick said: "It has not escaped our attention that the specific pairing we have postulated immediately suggests a possible copying mechanism for the genetic material."[9] Their ideas led to a further articulated theory that became known as the central dogma about genetics. The DNA is inherited, it undergoes duplication during mitosis so that the "copying mechanism" does its job. The double-stranded DNA also makes single-stranded RNA, which serves as a template for making proteins. That was the initial idea, but since the 1950s the central dogma has undergone considerable revision, in recognition of the many different kinds of RNA and other molecules.

Messenger RNA carries messages, and the transcription of DNA into RNA makes it possible to send the inherited template for the particular individual organism out into the cell. Then the RNA is said to undergo translation into proteins. Excellent histories of molecular genetics tell much more of this story, but the important concept here is that inheritance guides development. This suggests a kind of predeterminism, but many other factors intervene such that the process is highly epigenetic and environmentally directed as well. As François Jacob and Jacques Monod saw it in the 1960s, various forms of regulatory genes guide the process from heredity to development. Others have shown the influence of cell-cell interactions in shaping the environment and the developmental responses. A number of histories of molecular genetics set out different parts of this story, which is less about embryos and development and primarily about heredity.[10]

Thus, what had started in the 1950s to look like a "genes determine traits and therefore determine development" story became one in which the full complement of inherited genes provides a background that guides what follows. Other environmental factors also play important roles. The concept of phenotypic plasticity reflects this emphasis on the environmental roles; even individuals with the same genes can respond differently to the conditions in which their cells interact, the way the genes are transcribed or translated, and a host of other influences. Genes provide a starting point, but they do not dictate the result entirely. It has not been clear just how much determination and how much regulation is involved, but determining that balance has been the focus of considerable study ever since.

What the genes do is provide constraints. The realities of the physical processes of development also provide constraints on the range of what is possible in development. Conrad Hal Waddington introduced the idea of "canalization," based on the idea that a developing organism is like a ball rolling down a contour that includes channels. The constraints of the environment cause the ball to roll in some directions but do not allow it to roll in other directions unless something changes to give it a push over the hill between channels. This kind of developmental constraint restricts the future possibilities for development as an organism metaphorically rolls down the set of channels and makes choices (not consciously but physically). There are multiple possible pathways, but once the organism makes a choice it follows that path. In Waddington's view, the developmental constraints that keep organisms from redirecting are underpinned by genetic factors, seen as metaphorically pulling the channels downward (Figure 4.2). Genes determine development in this view, but even those most enthusiastic about genes have been forced in the last decades of the twentieth century to accept that it is really the genome as a whole that matters more than individual genes. The genes interact with each other and with other environmental influences.

Genes alone do not tell us much. This should not be at all surprising because developmental biologists have known this all along. They saw in the 1960s that genes are part of the story, but so are cell-cell interactions, the environment, and such factors as gradients and fields that are set up in the eggs and embryos from the very beginning. The phenomena of regeneration, as we have discussed before, suggests that genes are only a part of what causes development. What causes an earthworm or planarian to regenerate a lost part, for example? Is it an auxiliary set of genes, as Roux or Weismann might have suggested? Or is it some chemical and morphological response of the cells, as the developmental biologists largely held?

The recognition that the interaction of genes and development is more complex than a one-to-one causal connection led to the possibility that not all genes are equal, and to the idea of some genes as more active under particular conditions in leading to particular development responses, however those occur. Research in a number of laboratories explored this possibility and searched for "master control genes," so it was with considerable excitement that two groups announced the discovery of homeobox genes in 1983. These genes could explain a number of phenomena that had puzzled biologists. Walter Jakob Gehring at Basel, and Matthew Scott and Amy Weiner in Thomas Kaufman's group at Indiana University both identified a length of DNA sequence as having several important qualities.

This particular DNA sequence was conserved across widely different species, which suggested that it resisted mutation. When it

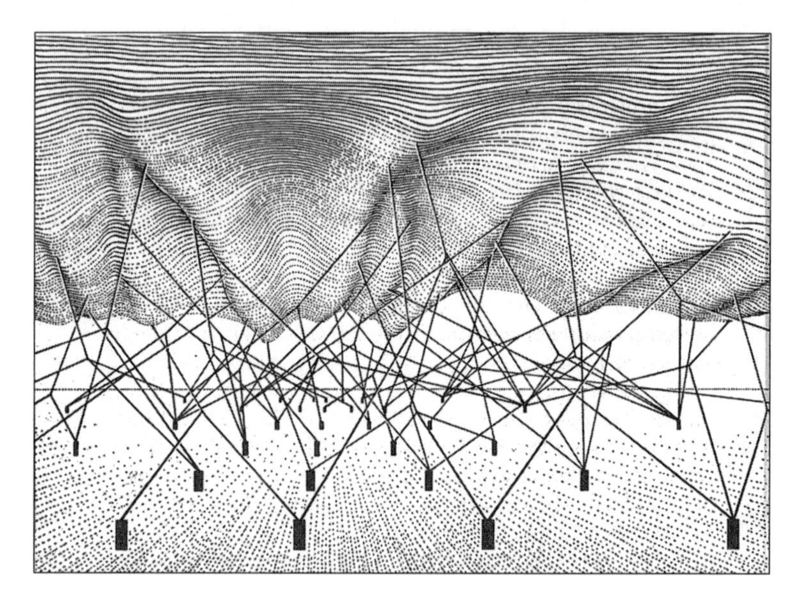

Figure 4.2. Epigenetic landscape model of Conrad Hal Waddington. Waddington viewed the genetic determinants as tying down and constraining the epigenetic landscape. That is, he saw the development of individual organisms in terms of the interaction of genes and the environment, which he called "epigenetics." The image here shows the genetic underpinning that causes that landscape of hills and valleys: the pegs represent genes, the guy ropes represent the chemical tendencies produced by the genes, and the hills model the epigenetic landscape, which is controlled by the pull of the ropes. (His most famous illustration showed balls rolling down channels; when the balls become stuck in the valleys, they are thereby "canalized.") From C. H. Waddington, *The Strategy of the Genes: A Discussion of Some Aspects of Theoretical Biology* (London: Ruskin House/George Allen and Unwin, 1957).

did undergo mutation, these homeotic mutants could transform whole parts of the body and change the patterns of development. An antenna could become a leg, in one of the early examples. The results suggested that there was indeed a master gene that controlled what the other genes do. Enthusiastic discussions of homedomains, homeobox, and hox genes helped provide a strong theoretical and experimental foundation to carry developmental genetics in new directions that brought together the molecular, genetic, and developmental. This work also brought in evolution and provided a foundation for evolutionary developmental biology, because of the widespread conservation of these special sequences.

New tools such as knockout techniques also made it possible to perform detailed studies of developmental genetics. These techniques allowed researchers to target particular genes, eliminate them, and then study the effects, supporting the claim that those missing genes caused those missing effects. In addition, genetic markers made it possible to observe changes in defined cells over time. Such techniques were first introduced in 1989 by Mario R. Capecchi, Martin Evans, and Oliver Smithies, for which they received the Nobel Prize in 2007.[11]

Nuclear Transplantation and Cloning

The emphasis on genetics and the nucleus, and even such work on cytoplasmic inheritance, raised questions about how to test the relative importance of their respective contributions to development. In 1938, Spemann had suggested the "fantastical experiment" of transplanting a nucleus from one embryo to another. We have no evidence that he actually tried to do so, but we do know that he would not have succeeded if he had. In 1952, Robert Briggs and Thomas King did succeed, though we do not have clear evidence that they knew of Spemann's suggestion and Briggs reported that he did not remember whether he had.

Nonetheless, Briggs and King performed the first successful cloning experiment through nuclear transplantation from one embryo to another, which was a logical extension of the transplantation research of Spemann and Harrison and also related to the studies of regulation that Driesch had inspired. They asked what would happen if they could transplant the nucleus from a donor to a host embryo: would the embryo develop like the donor or like the host? Such experiments would help get at the question about the relative contributions of nucleus and cytoplasm, and also about the ability to respond to changing environmental conditions.

Working at the Wistar Institute in Philadelphia, Pennsylvania, in the early 1950s, Briggs and King experimented with the leopard frog *Rana pipiens*. King reportedly performed the actual experimental manipulations, and Briggs worked on the theoretical understanding and the process. They started with eggs from which they had removed the nucleus. Today, this process is relatively easy, but getting to that point for these pioneers took a lot of trial and error as well as an unwavering conviction that it could work. They had to remove the jelly coat that surrounds the egg, then carefully reach into the egg with a very, very sharp glass needle and lift the nucleus out. When the technique is properly performed, the nucleus pops out, leaving the enucleated egg otherwise undamaged and full of cytoplasm.

Briggs and King then had to figure out how to remove an intact nucleus from another cell, a process that was more difficult because they did not want to damage the tiny structure of the nucleus. King developed a procedure using a pipette of just the right size: a little bit larger than the nucleus itself but smaller than the whole cell. This pipette was large enough to hold the entire intact nucleus when he sucked it up into the tube, and also large enough to break apart the surrounding cytoplasm so that he took only the nucleus, not the surrounding material.

Briggs and King experimented with which cells they could use as donors in this way, and early stages up to what is called the blastula stage seemed to work best. In fact, only these early stages actually worked in their experiments. They then took the nucleus from the donor frog and transplanted it into the enucleated host egg. Each step took patience and care, and out of a reported 197 enucleated eggs injected, they got at least some cleavage in 104 cases. Because they wanted to test the results at each stage, they dissected some and let others develop further. Only 15 went on to postgastrular development.[12] When they tried using cells from later stages as the donors of transplanted nuclei, however, they could not get the process to work at all, leading them to conclude that this transplantation could only work with the early stages. They hypothesized that embryos become more determined with each developmental stage, so they lose the capacity to adapt to changing conditions. This interpretation, in turn, led Briggs to conclude that differentiation must be preventing the foreign nucleus from being compatible with the host egg cell at the earliest stages. Yet Briggs and King saw as an open question how far the process could go, and also the additional question how different the donor and host could be before they became incompatible.

In the 1960s, from across the Atlantic where he was working at Oxford, John Gurdon took up the challenge to perform the process with later stages. The story goes that Gurdon's high school teacher wrote of the young man that "I believe Gurdon has ideas about becoming a scientist; on his present showing this is quite ridiculous," and went on to add that "if he can't learn simple biological facts he would have no chance of doing the work of a specialist, and it would be a sheer waste of time, both on his part and of those who would have to teach him."[13] Fortunately, Gurdon persisted with science and went to Oxford University, where he became intrigued by the idea of cloning frogs. In 1962, he published his results in successfully doing what Briggs and King had not been able to do and had become convinced was impossible.

Gurdon took nuclei from epithelial cells in adult frog intestines and transplanted them into a frog's egg from which the nucleus had been removed. Using the species of South African frog *Xenopus laevis,* he produced ten tadpoles that developed normally.[14] Through a series of experiments, he discovered that, as with Briggs and King's experiments, about 30 percent of the nuclei donated from the blastula stage could develop successfully to the tadpole stage. In addition, though the success rates declined with later stages, he could do better than 0 percent from those later stage embryos. From just-hatched tadpoles, extracted nuclei would work 6 percent of the time; even in swimming tadpoles, 3 percent of the donations could work. Perhaps the embryos become more differentiated and less plastic in their abilities to adapt, but they do not do so all at once.

Gurdon's discoveries were extremely important in showing that cells are not, in fact, irreversibly determined at each given stage. Rather, it seemed that reprogramming can occur, but it was not clear how that happens. He later moved to Cambridge, where he still performs active research in what is now the Gurdon Institute in a career distinguished for its long and continued creativity. In 2012, Gurdon shared the Nobel Prize for his research on cloning and reprogramming cells. Unlike many others, he did not make the assumption that using nuclei from later developmental stages could *not* initiate development but rather hypothesized that perhaps it could if he could just get the conditions right.

Gurdon's results opened tremendous possibilities for nuclear transfer at even later stages, and through the 1960s Robert G. McKinnell, Thomas J. King, and Marie A. Di Berardino succeeded in injecting enucleated eggs with nuclei from adult frog kidney carcinoma cells. They had hypothesized that such cancer cells might be more flexible and less determined than other adult cells. Then in 1966, Gurdon and V. Uehlinger used nuclei from tadpole intestinal cells. More and more "adult" cells seemed to be transferrable, which raised interesting new questions about the extent of the regulatory capacities of embryos. As more experiments continued, it seemed that the limits to transplanting nuclei were not intrinsic biological barriers but experimental limitations. Yet many remained skeptical about what would prove possible using nuclei from still later developmental stages because there was such a strong assumption that differentiated eggs cannot possibly have such capacities to de-differentiate back to early developmental stages. Most researchers also assumed that the capacity to adapt through cloning by nuclear transplantation (or, in other words, to undergo reprogramming that restarted the developmental process at a new point in the process) would only be possible in lower organisms.

Karl Illmensee's claim in 1979 that he had cloned three mice using four-day-old embryonic donors generated tremendous excitement. Yet because of the strong assumptions that it should not be possible to clone such "higher" organisms as mammals, including mice, and because everybody else who had tried had failed, many questioned his results. The debates that followed became heated and at times rather unpleasant, with accusations of fraud and misconduct. At the time, critics and supporters alike had difficulty determining just what Illmensee had actually accomplished, and even in retrospect when the results have been repeated with different techniques, some of the details remain unclear. It is likely that he accomplished what he claimed, and cloning mammals now has developed as a big, profitable agricultural business. This has been very exciting in the possibilities it suggests for being able to use cloning in even more ways. The possibilities have also frightened those who fear such experimental manipulation of life. We will return to this story later when we look at the cloned sheep Dolly and reactions to announcements of her birth.

The ability to perform cloning shows that the transplanted nuclear DNA does have a great deal of impact. Although Sonneborn and others may have been right that there is some role for cytoplasmic inheritance, and although inherited structures such as mitochondria that reside in the cytoplasm do also carry inherited DNA, the nuclear DNA clearly does play the central role in guiding gene expression and development. The environment—both inside and outside the organism—plays a role, but the nuclear DNA sets up the range of possibilities that can be realized. In a sense, the DNA carries the potential for the final result, more or less as Aristotle envisioned it, and the epigenetic developmental process and regulatory reactions such as those that Driesch observed for the whole embryo also plays an essential role. Boveri had productively focused closely on nuclei, and Driesch also provided an important perspective with his emphasis on regulation, as Wilson did with his focus on cells as fundamental units. The historical perspective shows the ways that divergent research fields can come together to transform the thinking of each field.

Evolved and Computed Embryos: Gene Regulatory Networks

Developmental biologists largely drew on the newfound knowledge of genetics combined with study of development, as we have seen. Only by the 1960s did serious efforts arise to add evolution to the mix and to understand embryos as being at the intersection of development, genetics, and evolution.

A few researchers focused on the question of how changes in the genetic composition are connected to changes in the observed phenotypic outcomes. The theoretical work of Eric Davidson and Roy J. Britten brought together embryonic development with evolution in this light.[15] In what has turned out to be a very influential and forward-looking paper in 1969 in the leading journal *Science,* Britten and Davidson provided an ingenious model for development, based on the idea of gene regulation connecting heredity and development. Their model was highly theoretical and as such was also strongly attacked by critics who chided the pair for their theoretical approach and essentially pronounced that there was no evidence for their claim. Yet Britten and Davidson laid out their ideas as a theory, saying that they did not know the exact mechanisms, but that it was time to put forward the hypotheses and try them out. Importantly, they integrated what was known at the time, especially in the emerging field of molecular biology where a great deal of evidence had accumulated about repetitive DNA and different types of RNA products.

Britten and Davidson offered a conceptual-logical framework that focused not on individual genes but on the whole genome and its regulatory properties within a modular and integrated system. Researchers studying embryos have tended to like concrete, empirically driven studies that look at real, solid, material things. Too much hypothesizing seemed to traditional developmental biologists too much like the speculative theories from the early twentieth century, ideas that had been rejected long before. So when yet another new model arrived, the developmental biology critics felt they could ignore this one as well.

What made this theory different, however, was the way it brought together different lines of research in a unifying framework that was testable, and it implied a conceptual shift away from focus on single genes to the whole genome and its interactions. They started with the understanding that cell division is controlled by regulation of gene activity, which raised the obvious question about how that occurs. What mechanisms direct this process? they asked. They took as given that differentiation in cells often depends on external signals, requires integration of a number of genes, seems to involve a much larger genome in cell types that are higher than bacteria, and involves a number of repetitive sequences. They asserted that "we propose a new set of regulatory mechanisms for the cells of higher organisms such that multiple changes in gene activity can result from a single initiatory event."[16] They followed with a model, often referred to as a battery-type model, with different circuits that can operate independently to a point but are integrated by other gene actions (Figure 4.3). The model was simple and beautifully clear, and its elegant emphasis on evolution in the context of development and heredity made it quite radical. At the time, some critics saw the Britten and Davidson approach as splitting the field of molecular biology, providing an alternative to the dominance by researchers characterizing individual genes in bacteria or bacteriophage.

Over time, Davidson and Britten have shown clearly that their model really works. The evolving model, which has shifted from early batteries to modules and integrated regulatory networks, ties together genetics, development, and draws on an evolutionary framework to account for the patterns that have emerged with further study. A lot of very close and careful study of purple sea urchins has demonstrated the effectiveness of their powerful concept of gene regulatory networks. Deeply embedded regulatory networks have been honed by adaptive responses to changes that occur during evolution. Or, as Britten and Davidson put it on the last page of their short but highly influential article in 1969, "It is therefore a requirement of a theory of genetic regulation that it supply a means of visualizing the process of evolution." And these networks drive development. Therefore, there is a developmental evolutionary approach that gets at patterns, causes, and the big picture.

The model and its subsequent empirical verification and expansion have gotten much farther than relying on what Davidson dismissed as just one phenomenological description of one embryo at a time, which embryologists technically had been doing. It has also supported a very simple and elegant theoretical formulation that triggers in many a response not unlike Thomas Huxley's when he first read Darwin's formulation of natural selection: "Of course, how stupid of me not to have thought of that!" The syllogism that emerges from Davidson's work is as follows: All phenotypes are the product of development. All phenotypic variation is therefore the product of some variation in development. Development is a highly regulated pro-

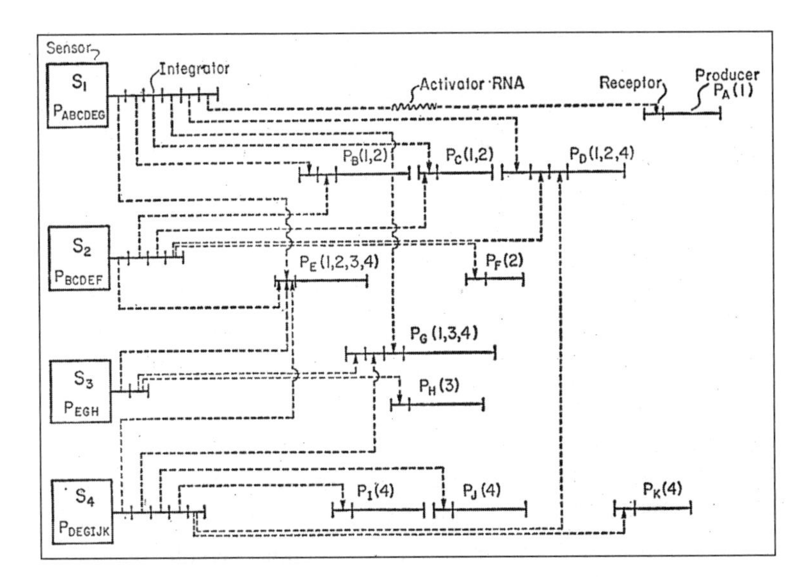

Figure 4.3. Gene regulatory networks model by Roy J. Britten and Eric Davidson. In this initial form, their model was based on overlapping batteries to represent genes and transcription. "The dotted lines symbolize the diffusion of activator RNA from its sites of synthesis, the integrator genes, to the receptor genes. The numbers in parentheses show which sensor genes control the transcription of the producer genes. At each sensor the battery of the producer genes activated by that sensor is listed." From Roy J. Britten and Eric Davidson, "Gene Regulatory Networks for Higher Cells: A Theory," *Science* (1969) 165: 349–357, illustration on p. 351. Reprinted with permission.

cess, where different genes are turned on and off in different cells. Understanding the regulation of these genes thus provides the basis for not only understanding development, but also understanding evolution.

Another interesting side to this story concerns the selection of what are called model organisms. As noted earlier, frogs have served as martyrs for science, and other species have played that role at other times. In the course of the twentieth century, researchers began to focus on selected organisms, arguing that they had so much information already invested that it made sense to continue concentrating on the same organism. They have argued that a particular selected organism provides a good proxy for human disease and therefore an effective model in that sense. In the United States, the National Institutes of Health first unofficially and then officially decided to declare some species as selected model organisms that would receive funding priority.[17] This approach has had its consequences. Who would ever have imagined that sea urchins—and especially a specific form of purple sea urchins—would become a model organism for studies of human health? Yet because of the work by Britten and Davidson, and the work of Rudolf and Elizabeth Raff and others, sea urchins have achieved that status.[18]

Computational Models of Embryos

As previously mentioned, one thing the idea of gene regulatory networks does is to move us from thinking about the embryo in terms of structure to a focus on underlying material causes. Davidson's approach has been to look at the network in terms of how each genetic effect affects all the others. The representation of the model looks like a complex circuit diagram, in which each "current" impacts the others to which it is connected. Sometimes the impact is positive and sometimes negative or inhibitory. This is all laid out with modules that are intricately connected. The connectedness provides a demonstration that only some kinds of changes are going to make possible evolutionary adaptations that work. Figuring out the system and its interactions involves a series of computational models that are sophisticated and coordinated.

Computational models also make it possible to understand interactions at multiple scales and to compare across different types of cells or divergent species. The term "computational embryology" began to pop up in various contexts starting in the 1970s, following on Britten and Davidson's thinking about regulation of development. More recently, such approaches have come to be called computational modeling. The British embryologist and theoretical biologist Lewis Wolpert offered one radical idea in this context: he has become famous for theoretical models that built on Alan Turing's 1940s attempts to explain patterns as a consequence of simple reaction-diffusion systems.

Wolpert introduced the so-called French flag model, demonstrating how clear and distinct patterns can emerge from the interactions of a small number of molecules. He then went on to ask the provocative question, "Is the embryo computable?" Wolpert's answer of "yes" has stimulated a lot of interesting work but has also drawn a great deal of criticism. The attention and lively debate has contributed to a shift of emphasis away from single experiments and toward a more integrated perspective.

Clearly, embryos are no longer just simple sets of cells that divide and then differentiate and gradually acquire form over time. Yes, embryos still develop gradually and therefore epigenetically; and yes, they still have cells and parts that differentiate and grow to yield morphological structure and physiological function. But computational biologists argue that understanding embryos requires a great deal more math and a lot more complex modeling of interacting complex systems than the earlier embryologists could have imagined.

This direction in biology emerges most emphatically in the context of the Human Genome Project. This effort, discussed in more detail later, brought together many different researchers in numerous countries to navigate their own political and economic environments in order to combine their data, the need for which resulted from sequencing in different parts and different examples of the human genome. Beyond individual genes, the emphasis was on the aggregated data from all the genes together. That is, all the genes from one individual can be aggregated and compared, but they can also be compared with the aggregated data from other organisms. Researchers initially had believed that when we got to that point of having the complete genomic information, we would be able to determine which genes cause which diseases. Personalized medicine would magically result, with inexpensive genomic results for each of us that would allow physicians to tailor treatments to our individual circumstances. That may happen, and it is already occurring in small ways.

Yet these medical applications of genetic data just emphasize the complexity of the whole. Not surprisingly, it turns out that we need not just genetic, not just genomic, and not just aggregated genomic data for comparison. We also need lots of other "omics" information. We need to know about the proteome—the aggregate of proteins that are correlated with particular genomic information. But between the genome and the proteome is the transcriptional information—which RNA are correlated in cases where particular genomic information is transcribed into particular proteins. The so-called transcriptome provides the aggregated data about transcriptions. Then there is the metabolome, which is even farther removed from the genome but represents the total metabolic results from particular genomes.

Big data come together in the National Institutes of Health's GenBank database, which is a repository for genetic information. Any data collected with public funding must be deposited there, and the mandate to contribute to this collection is already paying off for researchers. While geneticists gather more data and sequence more organisms, including whole genomes, others are performing computational analyses that draw on thousands of different sequences contributed by many different researchers. Aggregating data produces new challenges, of course, and requires new kinds of tools and new kinds of training. Those studying developmental phenomena this way are far from the cell-lineage observers or the experimental embryologists of the early twentieth century, and they are far beyond the modelers and theorists of the 1960s. Yet they are building on those rich traditions in ways that provide more and new kinds of information about development.

Nearly all who use such terms as "development," "genomics," and "computational biology" see themselves as doing something important, as getting at fundamental theoretical issues that have been largely unexplored or at least underexplored until now, and as allowing us at long last to begin understanding complex dynamic systems. Embryos are no longer merely sets of cells that develop and gradually acquire form over time.

Physical Modeling and Picturing Development

In addition to the computational models that continue to play an important role in stimulating new ideas, other kinds of models have also played a role. The Carnegie Collection of embryos began when Wilhelm His and Franklin Paine Mall collected the physical specimens, first in Germany and then in the United States. The Carnegie Institution of Washington supported Mall's work in collecting additional embryos and in studying the accumulating collection. Over time, the collection became home for a wide variety of serial sections, photographs, preserved physical specimens, and models as well as the embryos and fetuses preserved in glass jars. These models themselves became the foundations for teaching models, first made of wax, later made of plastic. The wax models proved very popular as well as useful, and historian Nick Hopwood has shown some of the ways in which the use of these models shaped thinking about embryos and about development.[19]

Imaging has always played a role in developmental biology. As we have seen, work at the end of the nineteenth and into the twentieth century very rarely included photographs and more commonly line drawings. By the 1960s, researchers were capturing developmental processes on videotape. Mount Holyoke College biologist Rachel Fink relates that the Marine Biological Laboratory at Woods Hole, Massachusetts, had movie nights, where the summer researchers would convene to show the films they had made in the laboratory during the winter of research back in their home institutions. The videos could show changes over time, and they could be stopped as needed to allow a closer look than was possible for one researcher peering through a microscope. Fink herself produced a marvelous set of films: *A Dozen Eggs: Time-Lapse Microscopy of Normal Development*. These begin with images taken while Fink worked with J. P. Trinkaus, and the series was sponsored by the Society for Developmental Biology.[20]

Video techniques with digital enhancement are a technology that has made possible amazing insights into what is happening in development. Throughout, embryologists and developmental biologists have worked closely with leading microscope companies and have depended on novel developments of these tools for seeing "inside" the developing embryo in new ways. By the 1980s, the many new tools had opened the way for different and complementary lines of research.

In the 1950s through 1980s, it was common to publish photographs of embryo sections illustrating what was happening with the cells. Yet obviously, it was not possible to photograph the hidden genes or evolutionary factors. The Gene Regulatory Network diagrams provided one approach to this issue; gene maps of various sorts provided another. It has been only more recently that researchers have been able to use digital imaging techniques to bring together serial sections of embryos with maps of genetic factors and also diagrams of the regulatory networks. It is worth looking at the rich variety of images now available on the Internet to see how all the pieces connect and how we can now "see" so much of what is not otherwise directly visible.

In fact, new tools for imaging morphogenesis have brought a return to earlier questions. Sophisticated imaging tools capture details that we cannot see in real time. After they collect these data, researchers can then use computational tools to provide detailed pictures of each step in morphogenesis. Other tools can help correlate changes in structure with changes in gene activity, which helps in understanding the regulatory actions involved in development. Computational approaches can then aggregate these data into an enhanced understanding of gene regulatory networks and their results. This work in progress promises to pull together a number of lines of earlier research, each of which had been limited in what it could do and in the connections it could draw.[21]

In addition, it is worth exploring the intersection of work on physical modeling and genetic modeling. Special techniques such as highly visible green fluorescent protein (GFP) have helped make modeling easier, for example. In the late 1960s and 1970s, several researchers developed tools for using GFP as a useful marker in living cells. That is, by inserting the gene that gives rise to this fluorescent protein (which is found naturally in some kinds of marine organisms and is traditionally taken from the jellyfish *Aequorea victoria*), it is possible to observe how the resulting now-glowing cells move. By the 1990, the technique had become much more widespread, thanks to cloning and additional technologies that have extended the reach into a wider variety of cell types and a wider range of organisms, and it is useful for showing visibly and clearly when a particular gene is being expressed.

The research using the protein is not unlike that performed earlier by Harrison and Spemann, when they transplanted bits of one frog to another frog that had very different pigmentation. In their case, they could then watch the differently colored tissue move to see how the cells differentiate through later developmental processes. In the case of GFP, the marker was inserted genetically rather than physically, but it has the same result of making it possible to see changes and to watch how one developmental step gives rise to the next.

We have seen that the hypothetical and observed embryo provided a starting point for understanding development, and then experimental embryology added new tools and new interpretations. Adding genetic and evolutionary perspectives as well as computational and imaging tools has brought so much more to the study of what was now called developmental biology. Yet nearly all of this work, aside from that Carnegie Collection of human embryos, was about animals other than humans. Much of the work was descriptive or comparative, but none was with humans—the human embryo was not visible in its natural or living state, not at any point in its developmental processes.

Until birth, human development takes place inside the mother and becomes visible only when the developing embryo (fetus) dies or when the mother dies; obviously such conditions were far from normal. Some efforts to visualize what was inside involved using X-rays, which provided a rough image of the structure of the hard parts; as we now know, X-rays also exposed the embryo or fetus and the mother both to often dangerous radiation. Sonograms, ultrasound, and tests such as amniocentesis (which involves extracting some cells from the fetus through a long needle to look at the chromosomes and genes) also give some idea of what is inside. Researchers have wanted to gather any information they can, but the embryos themselves have remained removed from direct observation.

Only in 1978 did the very first stages of normal human development become visible in living organisms via in vitro fertilization.

This technology took the egg and the developing embryo out of the woman and placed them in a glass dish for the earliest developmental stages, putting the embryo literally under the microscope. Only then did it become possible to compare the details and processes and stages of human embryos with those of other species. Those comparisons allowed researchers to understand much more about human development, and this has brought a rapid growth of research. However, the accumulating biological understanding and the fertilization technologies that resulted were also accompanied by increasing social and political repercussions.

5

The Visible Human Embryo

Within the context of changing theoretical interpretations and research to understand all kinds of animal embryos from many points of view, the human embryo finally became visible. As in other mammals, the developing human organism remains inside the mother until it is born. Historically, the only way to study it had been indirectly, until 1978 when the embryo came out, so to speak. As we know, researchers had seen early-stage human embryos before, as represented in the Carnegie collection, but these embryos were always dead, and many had been abnormal and spontaneously aborted. In 1978, thanks to the physician Patrick Steptoe and embryologist Robert Edwards, researchers figured out that the embryo could live in the laboratory for its very first stages and could then be implanted into the mother (Figure 5.1). In vitro fertilization (IVF) changed the lives of many—the women eager to have babies, the researchers eager for knowledge, and the public who now saw previously invisible, hypothetical embryos as visible and biologically real.

This chapter focuses on the innovations of the 1970s and their implications. It looks at fertility treatments such as IVF, contraception of several types, abortion politics, genetic engineering, and the legal contexts and responses. These themes overlap and intersect in

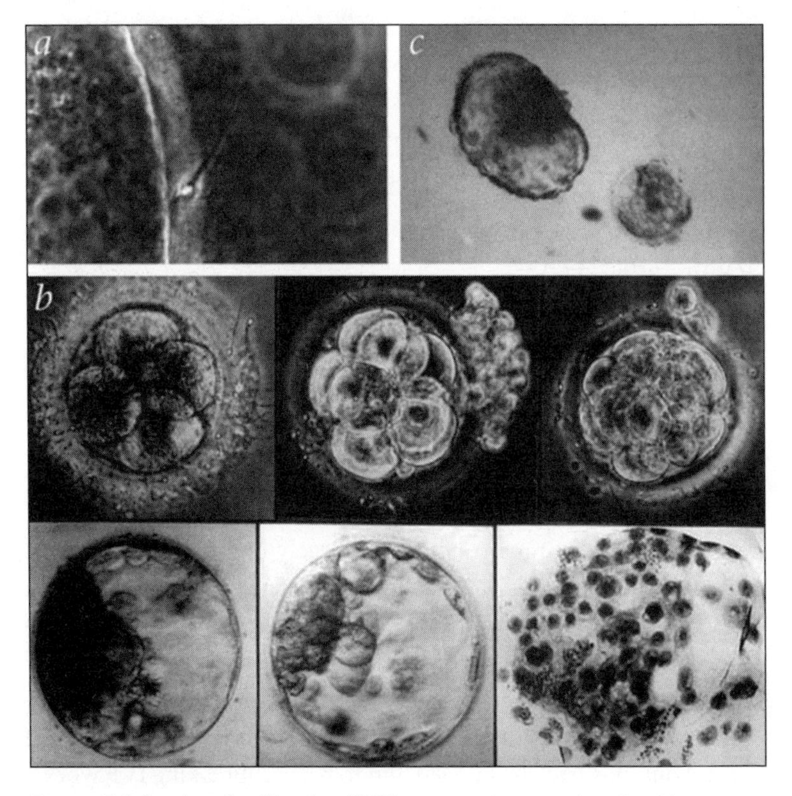

Figure 5.1. In vitro fertilization (IVF). From Robert G. Edwards, "The Bumpy Road to Human *In Vitro* Fertilization," *Nature Medicine* 7 no. 10 (2001). Reprinted with permission.

complex ways, and this is not the place to revisit stories about abortion politics, the women's movement, or contraception that have been told so well elsewhere. Rather, we will focus on the biological innovations and their implications for science and for social action. The story draws on selected episodes and examples among the many we could potentially discuss so as to point to key issues and questions and to note the articulation of competing ideas about the meaning of embryos and of their lives.

In Vitro Fertilization

Edwards, who received the Nobel Prize in 2010 for his work in fertility science and medicine, had long studied mouse embryos and stem cells without a major success. He had been born into a working class family in the United Kingdom in 1925, and he became fascinated with agriculture. After undistinguished studies initially, he applied to the University of Edinburgh to study animal genetics under Conrad Waddington, largely because a close friend had been accepted there. As a student under Waddington, Edwards began his study of mouse development, where he joined other researchers who as early as the 1950s had seen the importance of studying genetics alongside the morphological details of development. As his autobiographical account for the Nobel Prize committee laid out, this background led him to his pathbreaking IVF work.[1]

The 1950s was also the period when researchers worked out a classification system for human chromosomes, showing that each chromosome has an individual identity and role. They found that chromosomes retained their individuality and persisted through cell division and over time. It also began to be clear that some human diseases result because of abnormalities in the chromosomes, though the causes were not yet clear. Edwards took up the study of what happens in mouse embryos when there are one, three, or zero chromosomes of a particular type instead of the usual two—work that would begin to identify which features correlate with which chromosomes. Edwards began to work with a partner, Ruth Fowler, and they soon married and continued as a research team. One of their early challenges was to get female mice to produce a larger number of eggs at the same time so that they could control as many variables as possible in making experimental comparisons among the eggs. They used hormones to stimulate the mice to give up more eggs than they typically would in a normal ovulation cycle, a process called superovulation.

Perhaps most importantly, Edwards also mapped out the way the human cells go through a defined and timed sequence—which turned out to be crucial for making successful artificial fertilization possible. Human cells, Edwards determined, go through a complex process of dividing chromosomes and cytoplasm in regular, well-defined cycles, at a different rate from other species such as mice. From 1957 to 1958, Edwards visited the California Institute of Technology (Caltech) to study developmental biology, and there he became involved with studying reproduction, and in particular the related fields of contraception and immunology. When a sperm and egg join and then are implanted in a mother (in the case of mammals), questions always arise about whether there might be an immunological response between the embryo and mother. Something makes this process work in "natural" reproduction, at least in the successful cases. How, researchers wondered, could they get the timing and coordination right in the laboratory when they did not know much about the normal processes or why the timing mattered?

After he had returned to the United Kingdom, Edwards studied the release of eggs from the ovary and the possibility of culturing eggs in vitro. He also took up the study of embryonic stem cells, focusing on rabbit embryos. From this work, eight major papers resulted, but they did not immediately lead to follow-up studies by others, nor did they lead to any successes in other species. As his biography puts it, "That this work has largely been ignored by those in the stem cell field is probably mainly attributable to its being too far ahead of its time."[2] This happens often in science: a researcher has a brilliant idea but does not yet have the tools to conduct the research that would test the idea. In this case, it took almost two decades before similar work would be done in mice and excite tremendous attention.

Edwards took up other questions as well, such as how to cause fertilization to occur in the laboratory. He had worked out how to trigger the release of eggs, and he had even succeeded in obtaining a supply of human eggs in collaboration with partners at the Johns Hopkins University in Baltimore. But he was struggling with the problem of how to "capacitate" the sperm. This is a biochemical process by which the sperm gain the capacity to fertilize the egg. Without this step, fertilization does not occur. The step normally occurs when the sperm are already in a woman's uterus, so the question was how to cause the sperm to undergo the same process in the laboratory. Other researchers put together part of the answer, and Edwards worked out the details in humans: changing the alkalinity of the surroundings turned out to make the difference.

Then the physician Patrick Steptoe (1913–1988) persuaded Edwards to take up the problem of helping infertile couples have their own genetically connected babies. Steptoe was an obstetrician who had pioneered techniques for using tiny cameras inserted through small incisions to see inside the body. His technique, called laparoscopy, was already proving useful for other gynecological procedures, and it held great promise for successful extraction of eggs as well.

Steptoe and Edwards worked out a process of extracting eggs, collecting sperm, and fertilizing the eggs with the sperm "in the dish." In principle, this should not have been hard to do, they reasoned, but time after time the sperm failed to fertilize the eggs. The eggs did not divide or did not keep dividing. These experiments continued to fail until one trial succeeded: Edwards had accidentally gotten the cell cycle timing right, and the egg went from its relatively quiescent stage to its active stage at just the right time for the egg to be fertilized. He speculated that this attempt had succeeded because he had tried the fertilization in the middle of the night, as opposed to more normal hours during the day. As it turned out, that was when the cell cycles coordinated and fertilization occurred.[3] The importance of the highly regulated timing of the cell cycle has become apparent in many cases since.

Recounting the history of this discovery makes it sound as though one step logically led to the next, and as though making it all work was just a matter of time. In fact, the science did fall into place more easily in some ways than its acceptance. Edwards received rejections of grant applications, had difficulties in obtaining human eggs, and fielded ethical and political criticism. Fortunately for many couples who have benefitted from the processes they developed, neither Edwards nor Steptoe gave up in the face of opposition. They saw the clinical and scientific goals as worthy, so they persisted until they finally succeeded in 1978.

At the appropriate moment in its early development, the first fertilized egg was implanted into Lesley Brown in the United Kingdom for the first successful IVF in humans. Her daughter, Louise Brown, was born as a healthy child, and Louise herself later would have a perfectly normal, healthy baby as well. Edwards reported that he felt very close to Lesley Brown, who died June 6, 2012. Edwards himself died on April 10, 2013, not long after receiving the Nobel Prize in 2010. The IVF technique he had pioneered worked, seemed to produce normal babies, and quickly became a major medical enterprise in the United Kingdom and worldwide. For Edwards, as for Steptoe, being able to help patients become parents provided a powerful motivation for their work.

Human Development

Although IVF was developed as a fertility treatment, it also allowed the study of how human development works. This has led to a very detailed understanding of what happens in those earliest developmental stages up to what is called the blastocyst stage, at which point the embryo will die unless it is implanted into a mother or frozen for later implantation. The stages of development that can now be seen in the dish have filled in many details of the developmental process that had not been visible in the Carnegie collections. By the 1980s, it was becoming clear that the egg goes through a complex set of steps to prepare for fertilization, at which point a sperm cell joins the egg in fertilization. Because the process is complex and involves many changes, it is hard to point to a particular moment of "conception." It was also known, informed by other studies performed at the Carnegie Institution of Washington's Department of Embryology, that hormones help direct this process. Before, developmental biology and reproductive biology had represented two separate lines of research in many places; by the 1980s, they had begun to come together more effectively.[4]

After fertilization, the cell begins to divide. Cell lineage researchers have documented this process clearly for a variety of species, showing the differences within a species as well as the similarities that persist between species, in part dictated by the adaptations of evolution. In humans, the one cell divides into two, then into four, and then eight. What is intriguing is that these eight cells are all totipotent: as in other mammals like mice, the eight cells all have the potential to develop into whole individual organisms (that is, they retain their totipotency).

But with the next division, the cells seem to lose that total potency. In humans, after the eight-cell stage, the cells begin to divide at somewhat different rates; there is not a neat sequence of 16, 32, 64, and so on, but rather the numbers start to vary. Cells proliferate so that by the blastocyst stage, which occurs around day 5, the embryo has around 100 cells. The blastocyst has a layer of single cells arranged around a hollow ball that is partly filled with a mass of cells called (not surprisingly) the inner cell mass. That single layer makes up what will become the placenta that surrounds the embryo, and these are the only cells that are differentiated at this point. This set of cells is called the trophoblast, and it provides nutrients for the developing embryo and later fetus.

At this point, those cells that make up the inner cell mass are pluripotent. That is, they are plurally potent—they have the ability to become any type of cell among all the possibilities—but they do not have the capacity to become a whole organism, at least not without experimental intervention and then only rarely. The type of cell that each actually becomes depends on its location in the embryo and on what it eats (so to speak). The assumption is that if any of those pluripotent cells were cultured using the right culture medium, it could become any one of the types of cells that make up the body. It is obviously difficult to test a theory that says that a cell can become any one of all the possible kinds of cells because once the cell differentiates, we can no longer test whether it could have differentiated in some different way. The uncertainty problem aside, the assumption makes general sense, and there has been no evidence that these cells do not have the capacity to become any other kind of cell.

Thus, the defining features of these pluripotent stem cells are as follows. First, they can self-replicate, making more and more of themselves in cell lines (which are lineages of cells, cultured through multiple generations). Most cells begin to differentiate into more specialized types of cells rather than being purely self-replicating (having the ability to produce more cells like themselves), so this special ability is important. Second, given different culture media on which to grow, if they are pluripotent stem cells they can be caused to differentiate into any kind of cell. Cells inside the blastocyst remain pluripotent until the embryo is implanted and begins to undergo significant differentiation and growth.

Because the blastocyst is the last stage in humans where an embryo can be directly observed in the laboratory or fertility clinic's glass dish (because it must be implanted at that point to receive nourishment), this stage has practical importance for medical and research reasons. It is also the last stage in which the inner cell mass consists of undifferentiated pluripotent human embryonic stem cells and thereby has biological importance. The blastocyst stage in humans occurs at five to fourteen days, so that time period also has social significance, as captured in legislative and regulatory decisions.

All of this makes sense biologically and for practical reasons, yet putting embryos into glass dishes in fertility clinics did not immediately make embryos familiar to the public. Those who were not themselves seeking fertility treatments also needed to be able to visualize what is meant by an embryo and by developmental stages, and Lennart Nilsson helped provide that information.

Lennart Nilsson and the Public Image of the Embryo

Lennart Nilsson may well have single-handedly influenced the way society views embryos today. A Swedish photojournalist, Nilsson was fascinated by microscopes and the technique of microphotography and saw the potential for presenting the embryo and fetus inside the mother visually to a wider audience. To this end, he adapted medical laparoscopic techniques and used an endoscope to capture images inside the body. An endoscope is just a tiny camera, typically with a light source, on the end of a flexible tube. Unlike with typical laparoscopic procedures, endoscopes do not require an incision; they enter the body through natural rather than surgical openings. In a traditional colonoscopy, for example, physicians use an endoscope to see inside the colon.

Nilsson worked with physicians to use an endoscope to take pictures inside mothers' uteruses. Obviously, he did not do this without permission nor without reason—when an image was needed for medical reasons or was otherwise justifiable, Nilsson would capture an image, and eventually he showed the world his beautiful photography. He collected images of fetuses and some embryos at the early stages of development. His book *A Child Is Born* was initially published in 1965.[5] That same year, *Life* magazine featured Nilsson's work on the front cover, accompanied by a dramatic collection of photographs showing what most people had never seen or even imagined before. That special issue is said to have sold out of 8 million copies in just four days. The book has proceeded into further editions, each with more and more amazing photographical presentations of human development.[6] Nilsson made it possible for everybody to see—that is, to really visualize—the beginnings of human life.

Nilsson's widely published and much-discussed photographs shaped how people saw embryos (and fetuses), and they inspired the imagination about how the developmental processes might take place. We could all see these tiny forms dramatically appearing to suck a thumb, looking so innocent and deserving of protection. There is no doubt that these images caused some people to see the developing human as something to be protected and cherished, and many began to think of the developing organism as alive in ways that they had not really considered before. Nilsson's documentaries such as *The Miracle of Life* (1982) have reinforced that impression.[7]

In fact, in Nilsson's early work the fetuses he studied were sometimes dead specimens even though they looked alive. Also, although many have called them embryos, most of the specimens were from stages later than eight weeks, which would be properly known as fetuses. Regardless, what Nilsson shows in his images is clearly "true" in the sense that he has provided the photographs and does not intentionally mislead viewers to see something that is not there.

Yet he also is clearly presenting fetuses as perfect and formed from an early stage—he does not show physically deformed fetuses or those that miscarried because of problems, nor did he at first show the earliest stages, when the embryos are just blobs of cells in a bunch. His work does what brilliant art should do: it provokes us to think and to feel. However, the fact that his art presents human embryos and fetuses leads many to the impression that it is more than art, that it is science and medical fact as well—which it is, but it remains open to interpretation.

Critics point out that, especially in his early images, Nilsson focused on the developing embryo and fetus and largely left the mother out of the picture. Developmental biologist and historian Scott Gilbert has pointed out in a number of speeches and writings that the resulting impression is that a fetus floats independently of a mother, an autonomous individual needing nothing beyond itself.[8] The images do not all make clear that the fetus is actually attached by an umbilical cord to the mother and is surrounded by a placenta. Gilbert notes that Kubrick's classic movie *2001: A Space Odyssey* invokes a Nilsson-like fetus at the very end. The images had become so well known and so iconic that they came to symbolize starting a new life. These images of late stage fetuses have clearly helped shape the social impressions that embryos are essentially just little people.

Legal Context in the United Kingdom

As the embryo and fetus became more familiar and more visible to the larger public, and especially with the advent of IVF, many voices called for public discussion and for regulation. Some wanted protection for what they saw as innocent lives, and many researchers wanted clear rules so that they could perform research without being attacked the next day. Various protections arose in different countries for fetuses, especially in the later stages, but no clear shared understanding about what should happen with and for embryos has emerged.

With recognition of the biological importance of the blastocyst and the pluripotent embryonic stem cells within it, and in the setting of advancing IVF techniques and clinical hopes for stem cell therapies, the earliest regulations have thus focused on the blastocyst and on the fourteen-day period. In the United Kingdom and other countries that have regulated embryo research, the blastocyst has appeared to provide a logical end point for experimentation. In addition, the next stage brings formation of the beginning of what is called the primitive streak. Mammals (and birds, reptiles, and amphibians) experience a gathering of cells that begin to establish the lateral sides of the organism. This streak sets up the line around which bilateral symmetry develops, and it establishes the point at which gastrulation (when the embryo begins folding inward in a way that leads to the beginning of germ layer formation) will begin. This period when the blastula stage ends and the primitive streak begins, at roughly fourteen days in humans, is biologically critical. Legislation therefore often protects the embryo after fourteen days. In particular, legislation in the United Kingdom provides guidelines and a regulatory framework for the fertility business as well as embryo research.

After Louise Brown was born, public discussion began in earnest about IVF and its implications. In 1982, the U.K. government was the first to respond to the new technology by appointing a Committee of Inquiry into Human Fertilisation and Embryology to develop guiding principles for the control and regulation of the new reproductive technology. Mary Warnock, a philosopher, was appointed the committee chair, and she would file their thoughtful report in 1984. The committee included representatives of the government, social workers, physicians, scientists, and legal and religious representatives. Dame Warnock (as she later became), in submitting their report, acknowledged that "the task you set the Inquiry was not an easy one. The issues raised reflect fundamental moral, and often religious, questions that have taxed philosophers and others down the ages."[9] They recognized that infertility is a serious medical problem with psychological effects, that the treatment involves production of embryos in glass dishes before they are implanted into the mothers. Some of the embryos would be frozen, and others used for research. The committee carefully laid out its working assumptions and background considerations, then worked through the issues raised by fertility treatments, with the understanding that future research would develop in new directions.

One decision was to define an "embryo" as starting with the coming together of egg and sperm in fertilization. "We have regarded the embryonic stage to be the six weeks immediately following fertilisation which usually corresponds with the first eight weeks of gestation counted from the first day of the woman's last menstrual period."[10] Then the report addresses the medical problem of infertility, along with the full range of approaches to address the condition. Artificial insemination raised problems, they concluded, just as more recent technical innovations such as IVF. They considered both worthy of support, depending on the conditions warranting each, and thought both should be provided within the context of the National Health Service. Both egg donation and embryo donation should be allowed, under properly licensed and controlled conditions.

After considering what they saw as the currently realistic practices, the committee turned to human embryo research. Recounting the biology of the first fourteen days, and after laying out the arguments for and against allowing scientific research on human embryos, the committee concluded that "the embryo of the human species should be afforded some protection in law."[11] In particular, the research should be carefully regulated and performed only with appropriate licensing and oversight, and that "research may be carried out on any embryo resulting from in vitro fertilisation, whatever its provenance, up to the end of the fourteenth day after fertilisation, but subject to all other restrictions as may be imposed by the licensing body."[12] The committee called for the Parliament to pass legislation following their recommendations, and this led to the Human Fertilisation and Embryology Act of 1990, amended in 2008.

Contraception

Meanwhile, unlike the United Kingdom, the United States did not take up formal legislative discussion of infertility and reproduction. Rather, the discussions took place informally, in a variety of venues, among different audiences and contexts. Debates about reproduction focused primarily on birth control. Women had long known of mechanical practices for preventing contraception, though none has been consistently highly reliable. Intrauterine devices (IUDs), herbal concoctions, and various behaviors could either help prevent conception or terminate a pregnancy. The pharmaceutical industry added many more options, and the debates changed accordingly. In 1960, the U.S. Food and Drug Administration (FDA) approved the first oral steroid contraceptive pill, inspired by the hormonal studies of Carl Djerassi and developed by John Rock (a Roman Catholic) and the biologists Gregory Pincus and Min Chueh Chang working at the Worcester Foundation for Experimental Biology in Massachusetts. Inspired by the crusading work of Margaret Sanger, who had wanted women to have choices about reproduction and also wanted to provide birth control options for reducing population growth, Pincus had focused on finding an oral form of contraception. His textbook *The Control of Fertility* summarized the understanding of reproduction at the time and invited further study.[13] The implication was that much more was possible.

Researchers already knew about the role of ovaries, the balance of hormones including the estrogen family and progesterone, the cycles of reactions, the movement of eggs, and other features of reproduction. Studies of primates had helped provide significant information as well as new questions and theories. Researchers saw that contraception could work in any of several different ways: preventing release of eggs, or keeping sperm from reaching the egg, or blocking the egg from accepting the sperm, or halting cell division after fertilization, or preventing implantation of the fertilized egg in the uterus. Thus, chemical and mechanical barriers could work to prevent the entire process of conception from occurring. Drawing on research performed with other animals, Pincus worked out a successful chemical approach to contraception.

Condoms, which obviously provide a physical barrier, had been available since ancient times. In 1968, the federal government officially approved intrauterine devices (IUDs), which had existed for centuries and served as a barrier that prevented implantation of the embryo in the uterus. Other forms of birth control followed; in the context of the "sexual revolution" of the 1960s, a public perception emerged that the barn doors had opened for sexual activity, along with public expressions of the desire to control reproduction. Even U.S. president Richard Nixon spoke of the importance of family planning, and others emphasized the need to control population by making wise reproductive choices.

Of course birth control at the early stages to prevent fertilization, block implantation, or otherwise interfere with the process of implantation or gestation is very different from abortion. Abortion involves actually aborting—removing—the embryo or fetus (depending on the stage of development) and ending its development. To its critics, abortion involves killing the developing human. Yet we can look at those Carnegie developmental stages and recall that during the early stages, the characters that we would recognize as human are not at all fully developed. Only at the eight-week stage does the epigenetically emerging embryo become sufficiently formed that it gets renamed a fetus. Only after that point does it have even the rudiments of the essential organ and body systems. Only at twenty-four weeks does the fetus first become viable in the sense of being able to live on its own when removed from the mother—at least with our current technologies. Therefore, many have argued, the earlier developmental stages are fundamentally distinct from the later stages and should be treated differently than those later stages. If this distinction between early and late stages is maintained, then there is no compelling biological or secular reason to treat the two as equivalent.

Roe v. Wade

The United States, as mentioned earlier, has not passed any legislation regulating embryos, and as a result no federal guidelines exist for embryo research or clinics. Many supporters of the research and of fertility treatments are pleased by the lack of regulation because they assume that if the government did act it would restrict the possibilities. However, restrictions and ideally (perhaps idealistically) thoughtful regulation seem more likely as abortion politics enter many new arenas. Informed regulation can provide a very useful framework, especially in a context where there is considerable profit for business, high emotion among prospective parents, and many hopes and fears for clinical developments with stem cells.

How did the business of producing embryos through IVF and preventing them through contraception become connected to abortion politics? In part, this was an accident of timing. The United States has had a long history of strong political action against abortion, birth control, and other reproductive choices. The Comstock laws of the 1870s kicked off a century of efforts to restrict women's reproductive rights. Anthony Comstock, who served as a U.S. postal inspector, believed that sending birth control literature or devices through the U.S. mail was morally unacceptable, so he crusaded vigorously for legislative controls. By the 1870s, the federal government and nearly every state had enacted restrictive laws, and some states even restricted what information about contraception could be included in medical books.[14] The resulting climate led to the enactment of antiabortion laws as well. Interestingly, before this evangelical effort there had been relatively few such restrictions. In fact, that lack of long, sustained historical opposition would play a role in the *Roe v. Wade* decision in 1973.

Many commentators and historians have told the story of *Roe v. Wade,* and the case seems to provide a shining example of the triumph of civil rights (if you favor one side) or yet another step toward moral degradation (if you favor the other). The context tells us much about what was considered to be at stake at the time. The debates about the case occurred at the same time as the rise of the women's movement, which reinforced the desire of many women to control their own bodies and have the right to make their own choices about pregnancy. The idea that women should, in fact, have reproductive rights and choices gained momentum in the 1970s. As would occur later with IVF and the availability of new forms of conception, the availability of new forms of contraception contributed to major social change concerning reproduction.

In this social climate, the Supreme Court heard the case of *Roe v. Wade*. The case involved a single woman, Jane Roe, who had brought a class action suit to challenge the Texas laws prohibiting abortion. Henry Wade was the Texas district attorney who had defended the case. In the Court's decision on January 22, 1973, Justice Blackmun quoted the previous Justice Oliver Wendell Holmes in the case of *Lochner v. New York*: "[The Constitution] is made for people of fundamentally differing views, and the accident of our finding certain opinions natural and familiar or novel and even shocking ought not to conclude our judgment upon the question whether statutes embodying them conflict with the Constitution of the United States."[15]

The Justices reviewed the history of abortion and the reasons given for prohibiting it. One of the reasons involved prohibiting illicit sexual activity (though that was not one of the arguments in the particular case that the state of Texas had brought to the Supreme Court). A second reason might occur when the state finds the procedure dangerous and wants to protect its citizens—though this was not the case for legal abortions performed properly in 1973. Third, and more significantly, the state might have an interest in protecting the unborn life of a person, and perhaps it might even have a duty to do so. The latter reasoning hinges on the argument that the embryo or fetus is actually a person; but if so, the Fourteenth Amendment already provides protections. Or perhaps these earlier stages constituted only "potential persons." After considerable debate and careful study of precedents and arguments, the Court ruled that the definition of *person* "as used in the Fourteenth Amendment, does not include the unborn."[16]

Furthermore, in their decision, the Court concluded, "We need not resolve the difficult question of when life begins. When those trained in the respective disciplines of medicine, philosophy, and theology are unable to arrive at any consensus, the judiciary, at this point in the development of man's knowledge, is not in a position to speculate as to an answer."[17] As a result, they could only decide this case on the arguments presented, and this fact took them back to the merits of the reasons.

In fact, they noted, during the first three months (first trimester), pregnant women have a higher mortality rate than those who are not pregnant. Therefore, the procedure could not constitute a higher risk for the mother than proceeding with the pregnancy, and thus risk did not provide a reason to restrict abortion during this period. During the second three months or trimester, the fetus did not have rights because it was not yet viable. At this point, the interests of the mother should prevail. In the third and last trimester, however, states should have the authority to regulate and prohibit abortion.

At this point, they were persuaded, the fetus would be viable, and thus it might be a legitimate state interest to protect a potential citizen.

The *Roe v. Wade* ruling caused tremendous euphoria or anger, depending upon one's own interpretation of the meaning of life at all its developmental stages. The decades since have witnessed widespread continuing debate about appropriate abortion regulations, with especially vehement politicization in the United States. The states did respond to the ruling, though not immediately in most cases, and those responses took forms that represented the more liberal or conservative viewpoints of the majority of the state's constituents. As a result, we now have a patchwork of different regulations for each of the fifty states and other districts under U.S. government control. A number of states have enacted increasingly restrictive laws.

Yet as one social analyst has noted, in quoting an Australian reproductive practitioner, "abortion is like poverty: no one likes it, but it will always be with us."[18] He points out that, regardless of what many may wish, abortions happen in large numbers whether legally or not, for many complex reasons. The fact that abortions do occur and are found acceptable in some circumstances suggests that the Supreme Court and many people since then have accepted that developmental stages are, in fact, distinct and should be given differing legal and moral status.

The debates about abortion reflect very different ideas about the developmental stages and their meanings. The debates and their policy implementations involve discussions about what counts as abortion. Most people do not consider it an abortion when an egg is fertilized but prevented from implanting in a uterus, and there is clearly no abortion in any material sense when no fertilization has occurred, but this does not mean that there are not opponents to both those claims. While most people probably do not confuse or conflate the two issues of abortion and contraception, many do as we have seen in recent political debates. It is important to keep clearly separate those debates about contraception in the sense of *preventing* conception in the first place from debates about abortion, which include stopping a pregnancy that has already occurred.

New Forms of Contraception

Starting in the 1970s, building on the work of Alfred Yuzpe and others, and working in the context of emerging abortion politics, researchers developed safer products, and the pharmaceutical companies marketed hormonally based pills to be taken the morning after sexual intercourse. These emergency contraceptives, or "morning after" pills, became quite popular and were widely distributed, amid considerable controversy. Social conservatives rejected the idea that women should have the option for an easy way "out" after having sex by taking a simple pill. Women's rights advocates wanted women to have those options. And, of course, a compelling argument for emergency contraception is that sometimes women are raped and become pregnant against their will. In the United States, the FDA's decisions about this particular treatment, marketed as Plan B, became heavily politicized; nevertheless, its use was finally approved in 2009.

Thus, after the *Roe v. Wade* decision and over the years of interpreting what that decision means, the United States has settled into a somewhat uncomfortable set of compromises. The options available to women have become a patchwork of changing regulations at the state level. Some control is exerted through restrictions on sales, and some through legislation that allows pharmacists to exercise their individual values by refusing to fill orders for contraception. Those who oppose the use of any kind of contraception continue to seek ways to restrict the choices of women who want options. In 2012, debate even emerged over whether emergency contraceptives might properly be considered a form of abortion.

A *New York Times* feature in June 2012 on contraceptive treatments laid out the issues, revealing the climate of opinion at the time.[19] In particular, it addressed whether emergency contraceptives counted as abortion and thus fell under the restrictions imposed in various states—including requirements for parental consent, restrictions on who may fund their use, and other complex, abortion-related decisions. The 2012 presidential primaries brought the issue to the fore and were very instructive because they reflected decades of thinking on the topic of abortion as well as on what the government's proper role should be in controlling, or not controlling, women's bodies and reproduction. The 2012 public discussion also showed why debates at the intersection of science and society, especially those surrounding reproductive issues and embryos, are so difficult to resolve: even when the scientific facts are well grounded and accepted, the social interpretations are not equally clear.

In the case of the morning-after pill, the political issue hinges on biology—whether the drugs work by preventing fertilization or by preventing a fertilized egg from implanting in the woman's uterus. Opponents of the drugs, who included most of the candidates in the 2012 Republican primary, claimed that emergency contraceptives are "abortion pills," as presidential candidate Mitt Romney put it.[20] His staff found support for that claim in the packaging labels for the drugs, which stated that the pills work by blocking fertilized eggs from implantation. The news stories cited medical authorities such as National Institutes of Health and the Mayo Clinic's website as sources supporting the conservatives' viewpoint.

Yet, as the *New York Times* story in 2012 made clear to the public, the conservatives' claims did not accord with the best scientific knowledge of the day. In fact, researchers had definitively shown that the pills have other results. By July 2012, the Mayo Clinic had promptly updated their website to address the debate: "Morning-after pills do not end a pregnancy that has implanted. Depending on where you are in your menstrual cycle, morning-after pills may act by one or more of the following actions: delaying or preventing ovulation, blocking fertilization, or keeping a fertilized egg from implanting in the uterus. However, recent evidence strongly suggests that Plan B One-Step and Next Choice do not inhibit implantation. It's not clear if the same is true for Ella."[21]

The heated and oft-repeated claims of the political season did not fit with the current scientific facts. Following the *New York Times* feature, former Massachusetts governor and presidential candidate Mitt Romney modified his position and softened his attacks on the emergency contraception industry. He even held fundraising events with pharmaceutical company donors from companies that manufactured the pills. Yet the social conservatives who oppose abortion of any sort refused to accept the scientific evidence and persisted in trying to pressure Romney to continue his attacks. Even as the scientific evidence became overwhelming and universally accepted within the scientific community, conservatives refused to accept the evidence, accusing government authorities of caving in to "abortion advocates."

This episode, like so many others, shows clearly that some metaphysical beliefs are so strongly held that underlying faith and belief will trump any other evidence, including scientific. Some aspects of our public understanding of reproductive science, even including the definition of what counts as an embryo, are so contested that biology facts do not appear to matter. With such competing ideas about the meaning of life, it is wise for us to seek to understand the differences. It is also prudent to ensure that policy decisions are at the very least consistent with the best scientific evidence rather than in conflict with it. We will return to this issue in the context of stem cell research and its implications, but for now the focus remains on the driving question of how best to define an embryo and how to understand an embryo that is produced and modified in a clinical setting.

How Many Parents Make an Embryo?

We have said that an embryo starts with an egg and a sperm and fertilization. That is the most common course of events in nature, under normal circumstances: the egg comes from one woman, the sperm from one man, and the resulting embryo implants in the uterus of the woman from whom the egg came. This is also the most common circumstance in the clinic.

However, sometimes more "parents" are involved. Usually the egg is intact, as is the sperm (although we will read later of methods to enhance each), but the embryo that results need not be implanted into the woman who donated the egg; likewise, the sperm need not come from the man who will serve the role of father. So a child potentially can have many parents. The egg may be donated by one woman and the sperm from another man to produce an embryo that may be implanted and develop within a third person, the surrogate mother. Then at the time of birth, a different man and woman might assume the role of parents, usually those who paid for the procedure to take place. This means that the resulting child now has five different "parents" or contributors to the developmental process. Not only can eggs and sperm be donated, it is even possible to add mitochondria from one woman to another woman's egg to enhance it and make up for mitochondrial problems in the recipient. In each case, we are adding more parents in the sense of more contributors to the genetic makeup of the resulting embryo. The myth of the traditional family with two parents producing a child is too simple to cover such real-life cases.

The complexities lead to another question: is the gestating embryo and fetus dependent on the mother? That is, once the embryo is implanted into a woman's uterus and takes in food and eliminates waste through the umbilical cord, is it part of the mother or an autonomous individual? Obviously, both, to some extent. The U.S. courts in *Roe v. Wade* treated the embryo and fetus as part of the mother's body and thus as under the mother's control, at least to some extent. It is true that, without the mother, the embryo or fetus would die. Yet it is also true that the fetus in particular acts as a parasite. It is taking from the mother, and it triggers hormonal and other physical reactions in the mother during pregnancy. In a very real sense, the fetus is part of the mother and yet also an invader. The relationship is complex, and it is biological as well as socially and psychologically shaped.

Legal Responses to Changing Meanings of Parenthood

What have been the legal and political responses to the evolving understandings of human reproduction? Because the United States did not follow the U.K. example and work out clear regulatory guidance, what did happen? A mixture of different, often contradictory legal decisions have shown the lack of clear agreement about what embryos are and what we should do about them. Because medical matters remain in the hands of the state except where there is interstate commerce or other factors that trigger federal interest, we have a patchwork of conflicting decisions, and what is right in one state may be wrong in another.

Early cases dealing with IVF focused on who owned the embryos in question. For example, if a surrogate mother agreed to gestate an embryo and then decided that she wanted to keep the baby rather than give it to the party who had arranged for her surrogacy, that raised one kind of problem. Such cases have largely been settled by appeals to contract law, and they have followed the agreement in the contract. Some have attempted to declare the contracts invalid on the grounds that the embryos call for special protections, but these efforts have largely failed. Other cases have related to the ownership and control of embryos, and these too have largely been settled by contractual agreement, treating the embryos as property.

In cases when two parents who contributed the egg and sperm to produce embryos for later use then divorced, battles have resulted over who controls the frozen embryos; when a clear agreement was made at the time of fertilization, the agreement has been upheld. In cases without such agreements, some courts have treated the embryos as property; more recently, others have attempted to treat them as child custody cases. Because frozen embryos can be stored for years, over which time family conditions may change, such questions inevitably arise. For example, does the father own the embryos even after the death of the mother? Or do the mother's heirs now own them? Or if the mother is alive but has moved on to another relationship and no longer wants the embryos created with her former partner to be used, do her wishes not to have her eggs used by another woman trump his interest in having a child?

One actual example is the case of *Davis v. Davis* in 1992.[22] The Supreme Court of Tennessee upheld the claims of Junior Lewis Davis, who did not want the embryos that he and his wife had had fertilized in a fertility clinic to be used by his former wife, Mary Sue Davis. The couple had not established an agreement about the disposition of the embryos in case of divorce, and they divorced one year after the embryos had been fertilized and frozen. Because Mary Sue was infertile, she later sought to use the embryos, but Junior Lewis wanted to keep the embryos frozen. A trial court ruled for Mary Sue's right to use the embryos, deciding they should be treated as human beings and given to the mother to allow her to have a child. As the case continued through various courts on appeal, both parents remarried. Mary Sue wanted to donate the embryos to other couples, but he wanted them destroyed. The discussions through the court levels included debates about the nature of embryos or "pre-embryos" (as they were called in this stage before implantation), whether life begins at conception, and a range of other fraught issues. Ultimately, the state's supreme court ruled that Junior Lewis Davis's right not to be forced to become a parent won out over Mary Sue Davis's desire to donate the early-stage pre-embryos.

In contrast, in *Roman v. Roman* in 2006, the court in Texas upheld the contractual agreement that the couple would destroy their frozen embryos in the event of divorce.[23] The mother's desires to use them to have a child did not outweigh the previous contractual agreement. Other cases have reached a similar verdict, but as courts in some states have become more conservative and more willing to entertain the argument that embryos are more than mere property—and, in fact, have additional rights of their own—decisions could begin to shift in new directions.

The case of *Jeter v. Mayo Clinic Arizona*, from the Court of Appeals in Arizona, is instructive as it explicitly rejects the argument that embryos are persons in the sense that they could be considered to have suffered "wrongful death."[24] In this case, Belinda and William Jeter went through fertility treatments at the Mayo Clinic in Scottsdale, and ten cryopreserved embryos resulted. The Jeters then decided to transfer their treatment to a different facility and requested that the Mayo Clinic in Scottsdale, Arizona, move the embryos. Somehow only five of the ten actually made it to the new clinic, so the Jeters charged the Mayo Clinic with negligence. They claimed that the clinic had not only failed to protect and then lost their property, but also that the clinic had caused the wrongful death of the missing embryos. The claims about property and contractual agreements paralleled similar cases elsewhere, but the wrongful death claim was specific to Arizona's laws and the interpretation of them. The court determined that the embryos were, in fact, not persons, and thus there could be no wrongful death. Higher courts supported this conclusion and made clear that the frozen objects were pre-embryos. They were not implanted and were therefore incapable of developing further on their own—they were not viable, in this sense.

Decisions about when a developing embryo or fetus is considered a human with full protection under the law have evolved. Various attempts have been put forth, including a vote in Mississippi in November 2011, to define a fertilized egg as an embryo and as a person. The initiative sought to define a human being, deserving of full legal protections, as starting from the point of fertilization, cloning, or any other such procedure that resulted in a fertilized egg or embryo. The proposed amendment to the state's constitution would have protected all fertilized eggs. As one supporter from a group called Personhood USA put it, "The unborn child in the womb is scientifically proven to be a human being, and when it comes down to it we are a human-rights organization."[25] This is one of the arguments offered in favor of the claim that fertilized eggs should be protected.

This poorly planned attempt of personhood movement was defeated at the polls in Mississippi, and a similar outcome has occurred elsewhere. Yet considerable funding supports the measure, which has gained momentum in some states even in the wake of the volatile 2012 presidential election. Judging from the websites and speeches pushing legislation to protect personhood, the supporters care little about the actual science, but fortunately others do care. This movement shows once again why it is so important to understand the scientific debates, their social contexts, and how they have evolved over time.

The debates considered here pertain to the very earliest stages of life. As it becomes possible to keep premature fetuses alive from earlier and earlier stages, questions arise about just when a fetus is truly viable and whether that is the relevant criterion. We need to remain true to the science: if viability becomes possible much earlier than previously understood, then our policy should follow the biology. If it does turn out that the fetus is viable a little earlier than the last trimester, then perhaps abortion should be restricted accordingly. However, we do not know this yet, so there is no basis for arguments supporting the earlier restrictions on abortion rights.

The point here is simply that the science should inform the policy, whichever way it points us.

Under no definition are embryos viable on their own. They must be implanted in a uterus, and they must depend on a mother. To date, this has kept them in the realm of property where they are governed by contract law rather than considered full persons. It remains to be seen how such laws and interpretations develop, and how changing technological innovations will stimulate new responses.

Anthropologist Lynn Morgan, a professor at Mt. Holyoke College in Massachusetts, has provided an example of just how troubled our social understanding of development and reproduction has become.[26] She became interested in reports of a collection of fetuses at the college and went in search of the collection. She found some of the specimens but learned that much of the collection had been quietly discarded in the 1960s. To illustrate the human developmental sequence, researchers at this excellent women's college had diligently collected embryos and fetuses, just as other medical schools and biology departments were doing; however, faced with increased social tensions, the college had discarded most of its collection. Even serial sections of human embryos intended to educate about developmental diseases were quietly removed from public sight or thrown out. After Morgan uncovered and published this story, others found similar situations at their own institutions. The schools considered the social milieu too heated to maintain their scientific collections.

Genetic Engineering of Embryos

Beginning in the 1970s and 1980s, our understanding of the embryo and fetus as biological objects and as moral and social subjects has undergone considerable transformation. The physical structures and physiological processes of development also have come together with our evolving understanding of genetics. As the courts were deciding how embryos should be treated, and as women were deciding what they wanted for their own reproductive choices, research continued to raise new questions about the meaning of embryos from a biological perspective. Recombinant DNA in particular brought new challenges to traditional interpretations of how heredity works and how much is actually determined by inheritance, even in cases where genes clearly are a major factor driving development. What if we can change the genes? This is what recombinant DNA techniques do, allowing researchers to combine and recombine pieces of DNA from different cells, different individuals, and even different species.

It is striking how busy the 1970s were for reproductive medicine and biological science. Both IVF and abortion changed the social landscape considerably, altering the context in which reproduction took place. The science of recombinant DNA brought genetics into the picture in new ways. As the public became aware of the ability to recombine pieces of inherited material, debates muddied the waters of reproductive politics. Working with pieces of plasmid DNA in bacteria, as was happening in 1973, may not seem immediately relevant to abortion, but the fact that both were being highlighted in the news around the same time was not irrelevant.

In 1973, researchers announced that they had developed techniques for using a restriction enzyme to cut segments from DNA at precisely defined (or "restricted") sites. For this work, three of the original recombiners received the 1978 Nobel Prize for Physiology and Medicine. In public, the researchers emphasized the tremendous potential applications for manipulating pieces of DNA. Perhaps it would be possible to use this technique to add particular genes to patients with particular diseases caused by missing genes. Or maybe the technique could be used in the laboratory to cause bacteria to serve as factories of a sort, churning out such products as insulin. Indeed, this latter prospect soon proved realistic, which was very exciting.

At the same time, some of the researchers were nervous about the possibilities. It is all very well to combine known bits of DNA to achieve desired and known results, but we do not know everything. Some of the scientists worried that perhaps we needed to know more about how DNA combine before risking situations in which they might combine in undesired, dangerous ways. Two useful books, *Playing God* by June Goodfield and *Genetic Alchemy* by Sheldon Krimsky, present the debates and decisions of that time, including a discussion of the fallout from the important Asilomar conference in California on safety and recombinant DNA, which received tremendous public attention.[27]

In this episode, a group of scientists acted responsibly, and made every effort to protect the public and to ensure that they understood the risks of their research. Their efforts to be open and honest backfired in some respects, and led to a public outcry that was probably far more energetic than if they had just pretended that all was well. Maxine Singer, the retired director of the Carnegie Institution of Washington, commented on how she felt at a hearing in 1977 in Washington, D.C., at the National Academy of Sciences: "It is difficult to take demands for bans on 'all' recombinant DNA research seriously when they come from those who demand to be heard but do not stay to listen." Understanding that she had herself been one to raise questions, she said, "Those now labeled 'proponents' of the research worked long and hard for prohibiting certain experiments and matching containment requirements to estimated risks in others. The cautious analytical approach is a discouraging tactic against uninformed fear, mysticism, and political opportunism. But it must continue; nothing less than science itself is at stake."[28]

Singer called for scientists to remain cautious, analytical, and open, even when their opponents were not. This is a theme that recurs as developmental biology begins to produce engineered products. Small advances can bring vehement and angry opposition, often based on a lack of understanding. Scientists find it especially annoying when the lack of understanding is willful. They do not expect all members of the public to understand everything, but intentional misrepresentation and misunderstanding of the facts does, as Maxine Singer suggested, fly in the face of science itself.

The Public and Visible Embryo

The period of the visible human embryo inevitably made the embryo more public. Because it is visible and because it is a highly valued commodity, the embryo became a social as well as a biological object in new ways. This led to a patchwork of state decisions in legal cases about who owns embryos. When IVF produces embryos, typically it produces more than one individual or one couple can use. Sometimes these extra embryos are donated to other couples, which itself raises many complex questions about surrogate mothers and about who is related to the embryos under these special experimental conditions.

To review the key points, at first most courts ruled that embryos are property. They are a product of two individuals, and those individuals can retain or renounce their rights to their property. As surrogacy and donations became more common, we have had many new questions to address. As mentioned previously, embryos can come from donated eggs, donated sperm, donated fertilized eggs, or even donated wombs to gestate them. Families come in diverse forms, and so do their offspring. It is fairly clear what is going on biologically, at least in the early stages, so it is intriguing that society has accommodated and even embraced this range of options.

The typical embryo starts with inheritance from eggs and sperm passed on from parents, is the product of evolution, develops through a predictable set of stages, and responds to changing conditions in a regulatory way. All embryos do this, under normal circumstances without significant engineering, and we can figure out ways to make sense of these embryos. With recombinant DNA, however, we must address scientifically what actually happens and what it means when the situation is highly engineered through laboratory manipulations. In the next chapters, we will look at the implications for the biology and medical applications. First, we will look at one of the most highly discussed and feared social implications. The emphasis on inherited, predetermined development when combined with recombinant DNA technology brought concerns about whether genetic research could bring with it a new eugenics movement and whether genetic engineering was a new form of Nazi-like human experimentation.

Eugenics

For the first half of the twentieth century, advances in understanding and in controlling embryos and development remained modest. Biologists were busily learning all kinds of things, as we have seen, but they were largely accumulating knowledge and were not yet in the position to use it extensively. The more promising and more widely embraced approach to controlling life during this period came with eugenics, which was seen by its supporters as offering hope for engineering healthy populations. Even though the eugenics social movement did not focus on embryos and did not add to or draw on a biological understanding of embryos directly, the emphasis on controlling life contributed to the mindset of what Jacques Loeb called the "engineering ideal." Eugenics seemed to offer a way to control reproduction and thus, indirectly, the production of embryos of particular kinds.

Eugenics involved a Progressivist ideal of drawing on the science of Mendelian genetics to control life, with the goal of improving society through "good breeding." Because the human population included what looked to observers like clearly defective traits that seemed to run in families, why not adopt the public health stance of restricting the production of more babies from such families?[29] The idea of the perfect family, perfect children, and presumably the perfect society that would result attracted those who wanted to get rid of negative characteristics and also those who envisioned a positive society. Engineering reproduction in this way provided an attractive opportunity; the increasing availability of birth control information and devices contributed to making this imagined goal more realistic. The more radical step of embracing sterilization laws also seemed sensible to many, and a number of U.S. states passed laws to sterilize the "unfit" (the meaning of which varied). The estimates vary because of inadequate record keeping, but a cautious estimate holds that roughly 61,000 Americans had been sterilized by 1958, with over 20,000 of those in California.[30]

During the 1920s and 1930s, eugenics also played a well-documented role in guiding U.S. immigration restrictions and in informing social policies in Nazi Germany. By World War II, however, the biological understanding of developmental variation had called into question whether the crude sterilization strategies could really achieve the desired results for the population that the eugenicists had envisioned. In fact, it became clear to many biologists fairly early in the twentieth century that the scientific justification for eugenic action was weak, even though some remained enchanted with the possibility of controlling populations and thus improving public health and life. Engineering the population to improve it might offer a worthy goal, but by the 1930s it began to become clear that we simply did not know enough about what causes the expression of particular traits or about what is considered a desirable improvement.[31]

If policy followed scientific knowledge, surely eugenics would have lost all support. Yet not only did that not happen, but eugenics thinking has continued to reappear in a variety of ways, often under the guise of promoting public health and sometimes in the context of using genomics information to avoid "defects." The discovery of the structure and nature of DNA in the 1950s actually reinforced the decline of eugenics—for a while. The blunt instrument of sterilizing individuals in the hopes of preventing "defective" births did not fit well with the emerging biological understanding of the mechanisms of genetics. Only much more recently have researchers developed affordable, realistically useful tests for particular genetic traits, which have made it easier to interpret what look like genetically caused "problems." This has renewed hopes for genetic engineering to correct such problems. In looking at the hopes for genetics and genomics, it is worth keeping in mind that assumptions about genetic solutions have largely proven misguided or otherwise ineffective.

Genetics in the early twentieth century seemed to suggest that inherited genes guide or perhaps more directly determine an individual's traits. Yet genes do not act alone; the environment also plays a role in gene expression. Therefore, our phenotypic characteristics result from some mix of genetic and environmental factors. The dominant view of the early twentieth century recognized that heredity works together with the environment, at least to some extent, even while eugenicists hoped to control the nature side.[32] After the identification of chromosomes as made up of DNA, and the discovery of the double helix structure and of how DNA works, attention returned to the genes. What are genes, where are they, and how do they work? If only we could map the human genome, which is the set of all the genes of a particular individual, then we could start to discover which genes correlate with or perhaps even cause what traits. Or so it seemed.

A new kind of test appeared in 1989. Preimplantation genetic diagnosis (PGD) made it possible to use polymerase chain reaction (PCR) technology—which amplifies the effects of genes by allowing for minute samples to be replicated on a massive scale to get enough material to run genetic screenings—to test whether a cell carries selected genes or chromosomal abnormalities. Chromosomal abnormalities might come from abnormal divisions as well as genetic factors. As more genetic mutations are associated with diseases, PGD technology has made it possible to detect the conditions considered to be genetic diseases, with tremendous potential for identifying embryos considered to be at risk.

Previously, the only option had been to wait until the fetus was already partly developed, test with amniocentesis or other methods to detect genetic or chromosomal problems, then agonize over whether or not to abort the fetus. Most prospective parents would prefer to test an embryo before it is implanted and not to implant it if it has a clear abnormality or a defect that will lead to a known disease. By allowing such preimplantation diagnosis, PGD has offered far more options.

Of course, the very first cases of the test involved risk because it requires having something to test. The technology had been proven to work in animal models, notably mice, where it produced accurate test results, was safe, and gave rise to normal births. In humans, researchers take a cell from the embryo, often from the eight-cell stage, and test that cell. This assumes that the other seven cells would form a normal embryo.

The test has not been without controversy, bringing debates about which conditions parents should be allowed to test for. If we know there is a gene for something that is not considered a disease, is it still acceptable to test for it and choose only the embryos with that gene? If it were possible to perform such a test, would it be OK to test for blue eyes, or blond hair, or traits such as intelligence or athleticism in sports? What about deafness? It might seem acceptable to test for deafness to avoid it, some have argued, but what about when deaf parents want to have deaf children? Is that OK, too? There is still much room for debate, and a patchwork of practices and policies exist with no clear regulations or laws in the United States to guide such decisions. Organizations such as the American Society for Reproductive Medicine produce guidelines to fill the gap left by legislators.

The same kinds of issues hold for more recent testing methods, including noninvasive prenatal testing of the mother's blood to determine whether the fetal cells circulating there carry genetic "problems." New forms of testing promise earlier results, safer tests, and increasing ways of gaining knowledge about the embryo and fetus. Testing is not the same as controlling, of course. Having knowledge does not dictate how that knowledge will be used. Yet the kinds of knowledge we have and the range of ways it is possible to use it do have impact. In the following chapters, we will turn to explicit efforts to engineer and control life. Sometimes the focus is on individual life and engineering to improve individuals by affecting the embryo. Throughout any consideration of genetic testing runs the fact that we do not have a shared, accepted social understanding of what embryos are or of what we want them to be. Contested definitions raise challenges for social and individual decisions. What we can all do is to start from an informed understanding of what embryos have been thought to be, what has shaped those understandings, and which biological and medical factors were important, as understood by the best available science of the day.

6

The Idea of Engineered and Constructed Embryos

Engineering embryos to manipulate cells into what we want them to be or to do is an old idea. Only recently has the idea of constructing embryos out of nonembryonic materials or parts begun to seem possible. Yet recent ideas of constructing embryos and thereby organisms from scratch are an extension of suggestions about engineering and experimental manipulation already laid out much earlier. This chapter looks at the history of engineering attempts, when researchers first began to imagine that they could manipulate embryos to suit their interests.

The story starts with Jacques Loeb and his engineering ideas, then proceeds to look at disparate examples. The emphasis remains on the engineering efforts, especially the experimental manipulations in the late 1970s and 1980s that focused on research first, applications later. The following chapters will look more closely at examples of engineering for particular applied purposes. Here, we examine the context of cell culture, a necessary precursor for engineering work such as regeneration, creation of chimeras, cloning, stem cell research in mice, preimplantation genetic diagnosis, and the Human Genome Project. While looking at these scientific innovations, the discussion also will consider the ideas behind, motivations for, and some implications of wanting to do this kind of experimentation. This chapter lays out the idea and ideal of engineering embryos.

Gaining perspective on such thinking takes us back to Jacques Loeb and his mechanistic view of life, which was a philosophy that embraced the goal of engineering living systems.[1] His perspective was well ahead of most of his contemporaries, and his thinking stimulated imaginations at the time and ever since. (Google produces about the same number of hits for Loeb as for Thomas Hunt Morgan, and Morgan got a Nobel Prize while Loeb did not.) Loeb's ideas and vision for what biology is and could be have had significant impact. In the early twentieth century, several lines of research involved developing methods essentially to engineer organisms to do what the experimenter wants them to do. We discussed this goal in the context of experimental embryology in an earlier chapter, but this desire to engineer went beyond the research goals of manipulation or study—Loeb wanted to do both.

The Engineering Idea

Jacques Loeb's early years studying medicine in Germany were not terribly happy. He had trained as a physician because few other realistic career paths existed at the time for those fascinated by the sciences and especially for Jewish students. Loeb did not like the idea of practicing clinical medicine, but he saw no alternative as long as he remained in Germany. Fortunately for his career prospects, Loeb married an American, Anna Leonard, and they moved to the United States where they had a long, very happy marriage. In the United States, Loeb began to teach at the excellent Pennsylvania women's college Bryn Mawr, where fellow MBL researchers E. B. Wilson and T. H. Morgan had also begun their careers.

The story goes that the forward-thinking Bryn Mawr president, Martha Carey Thomas, was concerned when she learned that Loeb was Jewish, but she nonetheless offered him a place on the faculty so long as he made assurances he did not really feel or act religious, and his non-Jewish wife also was to keep him in line—meaning, presumably, that he knew how to behave properly in a Protestant American women's college. Today, an elite college's president expressing such concerns would be unacceptable; at the time, it indicated that Thomas was more liberal than most. Blatant, absolute exclusion of Jewish faculty members continued at many institutions up to the mid-twentieth century, notoriously including leading Ivy League universities such as Princeton. In contrast with those other institutions, Thomas allowed Loeb to join her faculty in spite of his background.

Loeb would spend his summers at the Marine Biological Laboratory (MBL) in Woods Hole, Massachusetts (Figure 6.1), together with the other hard-working researchers who made up that enclave, in what was reminiscent of a summer camp for doing what they loved.[2] In that atmosphere, big ideas such as imagining engineering life could take place without ridicule, accompanied by lively discussions among colleagues about how to make it happen and what the limitations might be.

In 1892, Loeb moved on from the small liberal arts college to the new research-oriented University of Chicago, where Charles Otis Whitman chaired the zoology department. Loeb already had been spending his summers at the MBL, which Whitman directed, so Whitman knew what he was getting when he invited Loeb to join him. Loeb enjoyed the opportunity that Chicago afforded to do more research and less teaching, but he struggled at times to understand what he viewed as a misguided enchantment with evolution among some of his Chicago colleagues. To Loeb, the direct material causes and action of life, as studied through physiology, mattered far more than some distant, inaccessible past. Loeb was fascinated by the idea of "controlling life," where a thorough knowledge of science promised great potential medical applications.

Figure 6.1. Jacques Loeb at the Marine Biological Laboratory. Loeb spent many summers at the MBL in Woods Hole, Massachusetts. With funding support from the Rockefeller Foundation, he worked in the laboratory facilities while also helping found and teach the MBL's physiology course. Undated photo, available through http://hpsrepository.mbl.edu/handle/10776/2179. Courtesy of the Marine Biological Laboratory Archives.

The medically oriented Rockefeller Foundation found Loeb's approach promising, and they provided funds for his laboratory at the MBL and supported his research for many years. The importance of such patronage was incalculable, as it made scientific research possible for those who were not themselves wealthy, as Robert Kohler explains in his *Partners in Science*.[3] Today, Loeb's name still graces the side of the newly renovated and enlarged Loeb Building at MBL, and innovative research and advanced studies continue to be produced from the MBL's famous laboratory-based courses.

Loeb wanted to engineer life in whatever ways possible. He focused on marine embryos because they were especially accessible, and he tried to understand how a functional, complex, integrated whole organism could arise out of a growing collection of increasingly differentiated parts. This central driving question for embryologists suggested to Loeb that if he could control and manipulate the conditions directing the earliest developmental stages, he could get at the underlying mechanisms directing life. He followed his counterparts at the *Stazione Zoologica* in Naples in selecting sea urchins for his primary study.

Sea urchin eggs are easy to collect, fertilize under controlled conditions, and observe closely to record the details and processes of cell division. As Boveri had shown, it is also relatively easy to break the wholes apart into pieces. This kind of manipulation had already helped Boveri address questions about how the parts and wholes fit together, and to assess the importance of the nucleus in particular. Driesch had used them as an experimental control to demonstrate and interpret the regulatory power of sea urchin eggs, as one cell could regenerate a whole embryo. Davidson and Britten have, since Loeb's time, shown how development, genetics, and evolution converge in these organisms. Sea urchins therefore give us numerous reasons for our continuing to favor them as experimental organisms for research.

Loeb discovered that he could cause eggs to begin developing on their own, without their having undergone the normal process of fertilization. As discussed earlier, via experimentally produced parthenogenesis, he could induce them to act as though they had been fertilized. According to the stories told by Loeb himself, his success was partly accidental. Unlike his colleagues, Loeb did not really like to go out and do his own collecting. Instead, he would order his specimens from the MBL supply department, and the collectors would set the buckets of whatever he had asked for in his laboratory. In one instance, Loeb moved his sea urchin eggs into a bucket that had a different solution of salt in the water than the usual, which was an accident in that he had not planned to experiment with varying the salt concentration.

An excellent experimental researcher, Loeb was opportunistic in the sense that he was prepared to observe anything out of the ordinary, especially when it departed from his expectations. Instead of throwing out what some might have regarded as a ruined bucket of specimens, he left them alone and discovered, to his delight, that the cells would divide anyway. In fact, the change in salt concentration seemed to stimulate the eggs to divide even without fertilization, in a process called parthenogenesis. The eggs then continued to develop, even without fertilization, all the way to the pluteus larval stage.[4] Loeb began to explore what factors would allow such parthenogenesis, including physical actions such as pricking the eggs or other chemical changes. It seemed clear to him that something internal to the egg was responding to the external conditions in a mechanical fashion.

The announcement of this newfound ability to "control life" generated tremendous public interest, and Loeb reportedly expressed dismay at some of the enthusiastic response. Some critics blasted him for what they saw as his efforts to play God with life, but others found the prospects for improving life fascinating. The turn of the twentieth century had brought with it a mood of progressive optimism, and a conviction that science and engineering could bring improvement to many arenas. Loeb provided the Progressives evidence that their hopes were well founded.

The *Chicago Sunday Tribune*'s headline on Sunday, November 19, 1899, enthused that "Science Nears the Secret of Life" and showed a picture of the MBL laboratories and the sea urchin, along with the exciting story of Loeb's discovery. As the subheading explained, "Scientist Jacques Loeb Develops Young Sea Urchins by Chemical Treatment; Discovery That Reproduction by This Means Is Possible a Long Step Towards Realizing the Dream of Biologists, 'to Create Life in a Test Tube.' " The headline in the *Boston Herald* read "Creation of Life; Startling Discovery of Prof. Loeb's Lower Animals Produced by Chemical Means; Process May Apply to Human Species; Immaculate Conception Explained; Wonderful Experiments Conducted at Woods Hole."[5]

As is evident from "Immaculate Conception Explained" and "To Create Life in a Test Tube," Loeb's work in 1899 aroused imaginations. Loeb insisted that he did not like all the attention and felt it was misguided. As with most others scientists of the day, he had no training in interacting effectively with the public.[6] As best he could, Loeb tried to explain the science carefully and not to exaggerate the implications of his work, but the press and some of his colleagues did that exaggerating for him. The tension between laboratory science and how it is translated to the public for different purposes continues today, especially when dealing with such dramatic topics as controlling, constructing, or tearing apart embryos through genetic engineering or stem cell research.

Loeb followed up his dramatic work on parthenogenesis with other studies of fertilization and mechanical changes in embryos, and he developed an approach that he called the "mechanistic conception of life."[7] The old debates about materialism and vitalism had largely given way to new debates; only a few stalwarts such as Driesch held on to ideas of vital "entelechies," which were not physical or chemical in the standard sense, to guide development. Nearly all biologists had embraced materialism and the sort of mechanism that Loeb called for, though most did not go so far as to write about the mechanistic conception of ethics, which Loeb felt should be added to the biological features of life. For Loeb, even our ethical behavior can be explained in terms of results from material causes and from our mechanistic physiological behavior. Loeb never accepted that religious or other special factors played a role in guiding human action. For Loeb, a firmly materialistic biology meant that he could work on engineering the living system and provide a vision that called for others to do so.

We have seen his early moves toward controlling, engineering, and eventually even constructing embryos. The engineering that Loeb envisioned started from a materialistic interpretation of life. By 1900, most serious biologists (other than Driesch, whose ideas we discussed earlier) thought of organisms in terms of material rather than special vitalistic forces or entities. They had cast aside ideas of spontaneous generation for mechanistic processes of fertilization and developmental processes. Some went further to emphasize that life is nothing more than the chemistry and physics of matter, organized in specialized ways. Stéphan Leduc, a French biologist, held such a view. His book on physical-chemical life and spontaneous generation in 1910 included ideas that have been cited as the earliest explicit endorsement for what would later be called synthetic biology.[8]

The ability to engineer embryos in particular requires a number of lines of research to come together. These would include the study of cell biology, cell and tissue engineering, transplantation and chimera creation, cloning as a form of transplantation, stem cell research in mice as the basis for stem cell transplantation, cloning through transplantation of nuclei, and the Human Genome Project to get at what were assumed to be the nuclear determinants. The underlying science in each case depends on continued study of the interactions of the nucleus and cytoplasm, genes and the environment, and heredity and development in the context of evolution. The Human Genome Project is one reflection of the desire to control life, and cloning provides an example of the intentional convergence of these lines of research.

Cell Biology and Cell Culture

E. B. Wilson, Oscar and Richard Hertwig, and others had done a fine job of laying out the basics of cell structure by the end of the nineteenth century, with continued additions into the twentieth century. Subsequent decades brought considerable understanding of cell specialization and function. Developmental genetics helped provide ways to interpret why the types of cells exist, how they differ, and what they do. Advances in cell biology brought many new tools for additional study. Meanwhile, the traditional cytological approach continued of preserving, sectioning, staining, and meticulously reassembling the pieces to construct a picture of cell structure.

New cytological techniques, such as those discussed in *General Cytology* and *Special Cytology*, two works produced by Edmund Cowdry, allowed more detailed study of the parts of cells and of different kinds of cells.[9] *General Cytology* included essays by most of the leading biologists of the time, who were working on understanding the chemical composition of cells and the actions of permeability as substances moved through the cells and across the membranes to the external environment. Cell reactivity to the environment was another important emphasis. By the 1920s, it was becoming clear that cells could not be considered mosaic tiles that aggregated with others to make up the whole organism. Rather, each cell was itself a dynamic, interactive whole, and their interactions raised many questions that Cowdry's contributors had begun to articulate. By 1960, *The Cell* by Jean Brachet and Alfred Mirsky would require six volumes to present the chemistry, physics, and morphology of the cell.[10] The collected work they reported reflected many new techniques and approaches to getting "inside" cells to understand the structures and processes.

By the 1940s, new techniques using electron microscopes combined with methods of fracturing the whole cell to see its parts more clearly had begun to reveal the finer details of the underlying structure. A few historians have discussed the work of these researchers, but this exciting period in cell biology deserves far more attention.[11] In addition, new ways to stain cells using not just dyes but also a growing set of biological markers revealed even more details. The growing army of researchers, supported by significantly increased funding from the government and private organizations, revealed not just the structure but also the functions and the cycles of the life of cells. Their research included examining how cells begin in the embryo as unspecified and homogenous, then undergo differentiation into different kinds of cells. Differentiation, growth, and morphological development go together, and John Tyler Bonner's study *Morphogenesis* provided one example of how the interaction happens.[12] His study of slime molds showed cells acting as individuals and aggregating to work together. Bonner wrote his book largely at the MBL in the laboratory of his Princeton colleague Conklin, and he reported having been inspired by the historical "aura" of the place.

As we have seen, Edwards and others were also discovering more about the nature of cell cycles. Cells undergo division, and the mitotic process involves many carefully timed steps in sequence. Cells divide, then divide again, and again, and again. This prompts questions about what happens as cells age and whether there is a limit to the number of divisions. Researchers also wondered whether the old model of cells dividing and then becoming specialized and remaining that way is really what happens. Is it possible that at least some kinds of cells actually undergo constant turnover and replacement throughout life? Do all cells do this, and if some do and some do not, why? And if cells are being replaced by others, how does that affect the integrity of the individual organism? What had seemed to be a defined organism made up of cells that divide and specialize and then stay that way was becoming a much more complex matter of interactions and feedback loops.

In addition, some researchers took up the question of how cells age. Most cells seem to have limits in the number of divisions they can go through, despite Alexis Carrel's ideas about immortality. In fact, in the 1960s Leonard Hayflick developed the concept of what became known as the Hayflick limit: normal cells in cell culture can divide only a limited number of times, at which point the cells experience a programmed cell death (called apoptosis).[13] Evolutionary adaptive factors probably make sure that cells do not become too old or survive too long; we know that aging can bring senescence and damage to the cells as well as to the organism as a whole.[14]

Hayflick came to his conclusions after study of aborted fetuses. A recent article in *Nature* shows that the work remains controversial even after fifty years.[15] In 1962, Hayflick was a microbiologist working at the Wistar Institute in Philadelphia. He wanted to study lines of cultured cells to understand how they change over time and what factors influence change. Taking lung tissue from a fetus that was aborted in Sweden, where the procedure was legal, Hayflick cultured a line of cells called WI-38 (for Wistar Institute, 38th strain). Hayflick's cultures turned out to be extremely productive: "Vaccines made using WI-38 cells have immunized hundreds of millions of people against rubella, rabies, adenovirus, polio, measles, chickenpox, and shingles."[16] In addition, the cells helped researchers understand development. Because the cells are considered normal rather than pathological or cancerous, they provided excellent controls for experiments as well. Yet the cells themselves had what some considered a controversial history, linked not only to the resistance by many in the public to using cells derived from human fetuses but also to the perplexing issues of ownership and rights to tissues and cells.

This episode is an excellent demonstration of the impact of the different meanings of embryos.

Much later work has showed that aging typically brings a reduction in the length of telomeres, the structures at the end of chromosomes. In 2009, Elizabeth H. Blackburn, Carol W. Greider, and Jack W. Szostak received the Nobel Prize in Physiology or Medicine for their work on telomeres and telomerase.[17] Telomeres are the strings of nucleotides that occur at the ends of chromosomes and in which sequences of nucleotides repeat multiple times. Under normal conditions, telomeres become shorter with each cell division, controlled by the enzyme telomerase, which obviously suggests that the repeated sequences are getting used up and might have a limited number of possible divisions. Blackburn, Greider, and Szostak showed how the process works and pointed to the role of telomeres in aging as well as in cell processes generally. It seems that normal developmental processes, cancer cells, and other mutated or damaged cells have different responses to changing conditions and over time, with different changes in the telomeres.

Tissue Engineering and Regenerative Biology

Knowing more about how cells work, along with improved techniques for cell and tissue culture, allowed research into tissue engineering, a term introduced at a 1987 meeting sponsored by the National Science Foundation. The founders of the field saw that they could bring together the principles of engineering and those of the life sciences with the goal of understanding how life works. They could then also work on developing materials to serve as effective substitutes for the biological tissue functions if the biological structures or functions failed. The engineered materials would not necessarily replace the same structure or even look like the original biological part, but the function would be achieved by the engineered replacement. This might include fairly simple examples like skin tissue cultured on a scaffold that helps it become the right shape. The role of scaffolding in development is just beginning to be understood because it requires understanding the biophysical structures in each step of development of an organ or part. Being able to engineer the organism to produce defined structures of the kind we want leads researchers to imagine a more distant goal of actually producing fully functional artificial organs. Even at the early stage in 1987, the leaders had a vision for what might become a kind of regenerative biology.

One leading early tissue engineer was Eugene Bell, after whom the MBL has named its recently established Eugene Bell Center for Regenerative Biology and Tissue Engineering. Bell showed the possibilities for translating fundamental scientific knowledge about regeneration into clinical medical applications through his own research and development of tissue engineering. His work reveals the cross-cutting approaches that helped establish the foundations for this important work.

Three decades ago, Bell and colleagues published a short but highly significant paper in *Science* titled "Living Tissue Formed in Vitro and Accepted as Skin-Equivalent of Full Thickness."[18] There, they expanded on earlier work that had showed the capacity of a lattice of collagen cells seeded with a kind of connective tissue cells called skin fibroblasts to develop into tissue of an apparently normal sort. That in vitro work led to the discovery, reported in the article, that they could graft living tissue on to wounded animals and that "the incorporated graft inhibits wound contraction so that the final area occupied by the graft is essentially unchanged." The study fell within a tradition of research on transplantation, tissue engineering, and wound repair, and it also pointed to potential applications for replacing tissue that had not yet been achieved with in vitro cultured tissues. The approach opened tremendous possibilities.

The authors noted the potential applications, "particularly since the substance can be cast into virtually any shape," but they cautioned that "its potential for reducing distortion and functional impairment in human injuries requiring skin replacement remains to be established."[19] Recognizing this need is the Bell Center at the MBL, where Loeb had first laid out his engineering vision. The Center offers the hope that "an understanding of tissue and organ regeneration in lower animals holds great promise for translating to medical treatments for serious human conditions, including spinal cord injury, diabetes, organ failure, and degenerative neural diseases such as Alzheimer's."[20]

Engineering and manipulation cannot themselves cause the tissue to serve as a skin (or other tissue) replacement for all the functions it needs to replace. At least, with tissue engineering, replacing structure is much easier than reproducing or replacing function. Understanding and replacing function requires considerable knowledge of the underlying biology, which takes time. The lag in realizing the importance of novel ideas is surely one reason why Bell's 1981 *Science* paper had been cited nearly 500 times by the beginning of 2011, with far more of the citations in recent years. In fact, the pattern of citations shows a continued interest from a wide range of researchers, with increased attention associated with the rise of interest in stem cell research and regenerative medicine. Bell's approach to engineering is grounded in biology, just as Harrison's work on wound repair or Morgan's studies of regeneration began with fundamental biological questions that led to broader implications.

Transplantation and Chimeras

Meanwhile, others have pursued research based on the sort of transplantation techniques that Spemann and Harrison first used so effectively to understand normal development. The approaches have been used to explore what engineering might be possible by moving bits of living organisms from one to another.

In England, Charles E. Ford and colleagues focused on mouse regeneration. They reported in 1956 on experiments taking hematopoietic cells (cells that give rise to the parts of the blood) from a normal mouse and transplanting those cells to another mouse that had been lethally injured by radiation. The recipient mouse recovered. Why did this happen and how? Could the transplanted donor cells have been taken up by the host and incorporated into a new individual organism? If so, the researchers had produced a new kind of chimera. Ford's group reclaimed the term to capture their exciting new science: "Although the very term chimera points to the antiquity of the idea, it is believed that the experiment reported here provides the first decisive evidence in animals that normal cells of one species may, in special circumstances, not merely survive and multiply in another, but even replace the corresponding cells of the host and take over their functions."[21]

Beatrice Mintz from the Institute for Cancer Research in Philadelphia took the approach further. In the 1960s, she improved the techniques for harvesting cells from different mice and "shoving them together," as she sometimes described the technique. She explained at the American Society of Zoologists 1962 summer meeting that "a method has been developed whereby some or all of the blastomeres from two mouse embryos in early cleavage stages may be readily combined: these cells, or whole eggs, quickly re-assort to form a single embryo which continues normal development. The best survival is obtained, and the simplest procedures required, when entire eggs are united at approximately the 8-cell stage."[22] Mintz discovered that making the procedure work required removing the zona pellucida (the membrane that surrounds the egg cell), and she developed a method for that removal that did not harm the egg itself. The cells from the different organisms combined together in about an hour, and the mosaic, as she called it, could occur by the combination of normal embryos or even of lethal mutants (Figure 6.2). Her research addressed questions about the relative flexibility

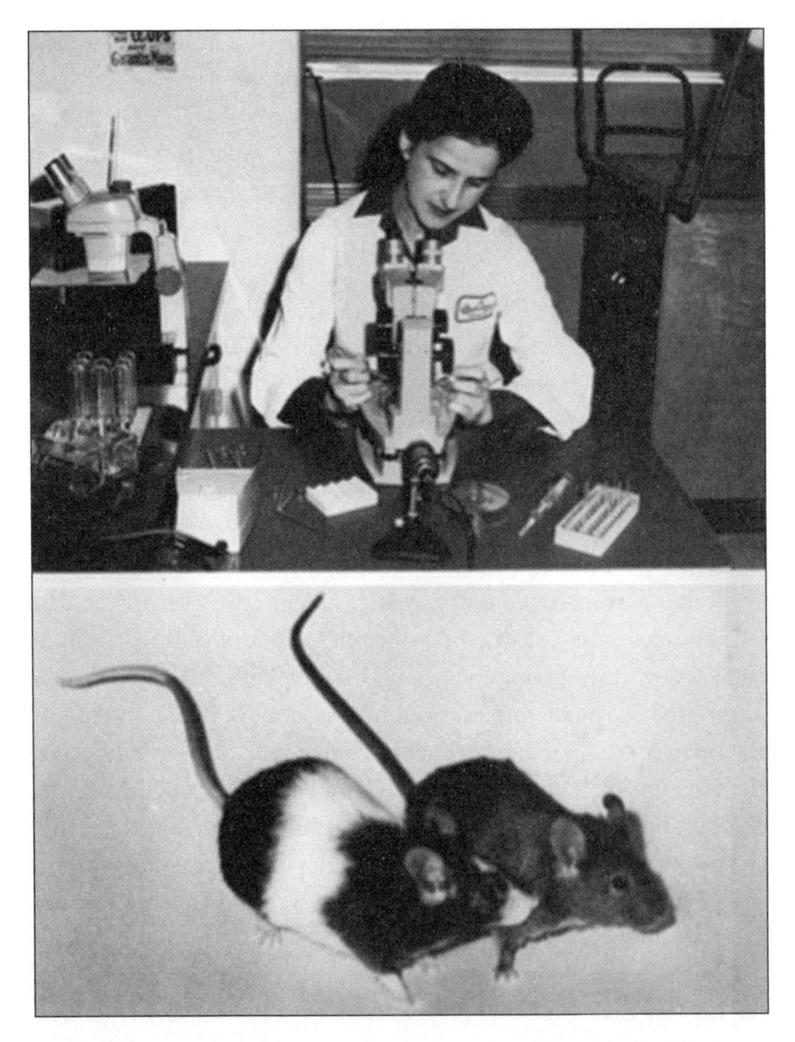

Figure 6.2. Beatrice Mintz and her chimeric mice. Mintz showed that combining cells or embryos at early stages could produce chimeric mice that appear normal in all outward respects. Photo available through the Smithsonian Institution, Acc. 90-105 - Science Service, Records, 1920s–1970s, Smithsonian Institution Archives, http://www.flickr.com/photos/smithsonian/6891505741/. Used with permission.

of individual cells in embryos as well as the way in which the individuals make up a defined whole out of separate parts.

By the 1970s and 1980s, the French embryologist Nicole Le Douarin had adapted and expanded the use of similar transplantation techniques to produce chimeras from different species and to ask a different set of developmental questions.[23] She and her colleagues wanted to track the fate of different cells as they contributed to development. Le Douarin combined cells from chicks and quails, and then ducks and chicks, and asked whether the resulting development of the neural crest in particular followed patterns like those in normal embryos, or whether they would become abnormal because of the unusual mix. The result, she concluded, was essentially normal. This again spoke to the tremendous capacity of an organism to incorporate parts and regulate them into a developing and functioning whole.

From our perspective today, it might seem rather surprising that Mintz's successful construction of embryos out of cells from different eggs or early preimplantation embryos and Le Douarin's creation of chick-quail chimeras did not raise more public interest at the time. After all, the experiments carried implicit possibilities for the creation of human chimeras. The mice that resulted from the combination of differently pigmented embryos put together from cells with different pigmentation developed stripes—they represented a very visible genetic mosaic, and were obviously quite impressive and photogenic.

More research along these lines performed by Mario Capecchi at the University of Utah and Oliver Smithies at the University of Wisconsin involved combining genetic material from one mouse into another as a genetic chimera. They found that the resulting healthy hybrid expressed genes from both parents. The two received the 2007 Nobel Prize for their work, which helped make such research visible to the public at last.[24] Before that time, this line of research largely escaped the attention of bioethicists, presumably in large part because the research concerned mice rather than humans and also because the field of bioethics had focused on clinical medicine and had not yet extended its purview to research by the 1960s.

More recently, bioethics debates have focused especially on chimeras made from combining human and nonhuman cells. The announcement that such "unnatural" combinations might be possible immediately raised concerns because of the interest in using such chimeras to test possible medical applications. Traditional medical protocols and regulations require that any proposed medical device or drug be extensively tested to be sure that it is safe, and in cases where there is risk involved that it is at least as successful as other options. This approach necessarily requires eventual testing in humans, but the standard research approach starts with testing in animals before taking risks with humans. We treat animals as a proxy for humans as a first step in testing safety of proposed treatments. With this in mind, it becomes more obvious why many would be interested in mixing human cells with those of other animals. This also applies to stem cells research: according to the standard protocols, it would be necessary to test human stem cells and human embryonic stem cells in other animals first, which requires mixing human and nonhuman cells to form a kind of chimera.

Immediately, opponents argued that it would be unethical as well as unwise to transplant human cells into animals.[25] Furthermore, it would also be unethical to test the cells in humans when we know so little about how the cells behave and when the results are so high risk. And just as the regulatory protections for human subjects do not allow researchers to inject humans with cancers or known toxins, presumably humans should not be subjected to experimental testing by introducing stem cells into human patients. These two objections led to much debate, with all sides seizing on the uncertainties to argue for their preferred positions. The prospects for transplanting transgenic cells—that is, cells from one species to another—have raised questions about the appropriate limits on what types of cell transplantations to allow. Ought researchers to put human nerve cells into mice, for example? Or human egg or sperm cells? Are some kinds of cells so uniquely human that we should never use them in such experiments? We have not resolved those issues and do not have clear answers.

In the United States, as with other embryo research we have discussed, no federal law guides what kinds of research are allowed. This leaves it to biologists to decide, bioethicists to argue, and individual states to regulate on their own. In the United Kingdom and Canada, legislative interpretations do provide a useful guide, such as the 2007 review "Hybrids and Chimeras" by the British Human Fertilisation and Embryology Authority.[26] This document lays out the circumstances in which combining cells from humans and other animals makes sense, and it acknowledges that sometimes such combinations will prove valuable and even necessary.

Such chimera research is occurring now, under some regulatory guidance in countries such as the United Kingdom, with valuable new discoveries as the likely outcomes. In Tokyo, an advisory ethics panel recommended removing the ban on combining human and other animal cells, at least for some kinds of experiments.[27] In both Japan and the United Kingdom, researchers are working on models and early experimentation to create chimeras from human and pig or human and mouse cells.[28] Researchers would like the United States to approve such research as well; however, although this type of research is not explicitly prohibited in the United States, any research with human cells must pass through institutional regulatory protocols. It remains to be seen what sort of impact the questions about ethics and the social interpretations of combining cells will have.[29] For example, those who believe that an individual's life begins with fertilization must surely regard transplanting cells and tissues to create chimeras as questioning what it means to be an individual. And increasing evidence has demonstrated that, under rare conditions, chimeras result not just in the laboratory but also in nature; even in humans, two early-stage embryos can fuse together and combine but go on to grow up normally. This natural capacity and the experimental ability to separate and combine cells at later stages raise ethical as well as practical questions, as bioethicists have begun to recognize.

Cloning

Another kind of transplantation involves cloning. Successful cloning of mammalian cells began in 1997, when a team of agricultural researchers led by Ian Wilmut in Scotland announced the birth of Dolly, the cloned sheep.[30] Their report, which appeared in the journal *Nature,* related the efforts by the group of agricultural scientists. Dolly's birth began not with fertilization but with nuclear transfer—here, the researchers had transplanted not just a nucleus but an entire somatic cell from an adult sheep. That is, they had taken a cell from an adult mammal and had produced a genetic clone of that donor cell. Wilmut became the visible spokesman for the research, though he explained later that his colleague Keith Campbell had provided the lead in making the technique actually work.[31]

Technically, cloning Dolly happened much as with Briggs and King's, and Gurdon's earlier cloning of frogs, by using transplanted nuclei—but with a twist. In the case of the frogs, the researchers had begun with an egg from which the nucleus had been removed; they then replaced another nucleus into that enucleated host cell. In the case of Dolly, they placed the whole cell into the egg, and in particular they used the cell from a mammary gland of the donor sheep. This meant that the cloned sheep Dolly was genetically almost identical to the adult donor of the mammary cell (with the addition of the little bits of cytoplasmic and mitochondrial DNA as well as the nucleus). The fact that they had successfully cloned an adult mammal excited the public. Did this mean that researchers could take cells from any one of us and generate an offspring genetically identical but delayed by a generation?

The news media went wild over this story. The report appeared in the very respectable scientific journal *Nature*. Typically, when a scientific report seems enticing, the news media honor a news embargo for a short period of time to give reporters a chance to do some background research and prepare their stories; the Dolly story initially had an embargo, but the British journalist Robin McKie broke it, claiming he had gone to original sources and had not relied on the *Nature* article.[32] The scientists judged that this journalist had actually produced a carefully researched and balanced story, but according to the journalistic conventions, the appearance of his article removed the restrictions on further reporting. If one journalist runs with an article before the embargo ends, they all can—a rush to publish ensued.

Gina Kolata reported the accomplishment for the *New York Times* on February 23, 1997, which broke the story in the United States. The first news report often gets picked up and repeated over and over, usually with little additional interpretation or even understanding because few local newspapers have science reporters. Kolata's Sunday edition headline read, "Scientist Reports First Cloning Ever of Adult Mammal."[33] Although Dolly was a sheep, the implications seemed sensational, as though researchers needed just a few steps to do the same with humans. Many of those reading the articles or watching the subsequent television news reiterations imagined a sole scientist performing the cloning of babies from adult humans.

Kolata also interviewed a geneticist from Princeton University, Lee Silver, who asserted his astonishment at such cloning: "It's unbelievable. It basically means that there are no limits. It means all of science fiction is true. They said it could never be done and now here it is, done before the year 2000." The way the *New York Times* highlighted the exotic implications of the story seems questionable, but

Kolata explained at a recent conference that for her science reporting is just another form of entertainment, and she does not feel responsible for the impact a story has.[34] In retrospect, Silver should probably not have made his glib, exaggerated comment on a weekend to somebody he knew to be a reporter for the *New York Times*. As a geneticist, he did not specialize in developmental biology, and as a Princeton University scientist, he may well not have been aware of ongoing studies in the applied agricultural world; if he had been, he would have found the cloning announcement less startling.[35] Yet nothing can quite excuse his claim that "all of science fiction is true." Clearly, Silver assumed that producing genetic duplicates must remain *only* science fiction because he had accepted the two basic assumptions of developmental biology: first, that differentiation and development go in only one direction, and that after cells have differentiated (in this case as mammary cells), they cannot be undifferentiated or redifferentiated into something else; second, that such later, adult, cells have gone too far along in development to remain flexible enough to adapt. Both assumptions turned out to be wrong.

As noted, public presentation of this story churned up a tremendous amount of excitement, including a great deal of worry and opposition. When the details became known, it turned out that a few years earlier, in 1995, Wilmut's group had already hollowed the nuclei out of many eggs in preparation for a transfer from donor cells that had differentiated but not yet become adult cells. In that case, they transferred 244 nuclei to the eggs, of which thirty-four embryos developed enough to be transferred to the uterus of surrogate mothers. These thirty-four produced five live lambs, of which two, Megan and Morag, survived to adulthood. These two sheep were the first mammals ever cloned from cells that had already been differentiated beyond the earliest stages.[36]

Yet their earlier results, published in scientific journals, did not receive much media attention. Wilmut and Campbell's team continued their cloning work, trying with nuclei transferred from donor embryonic cells, from fetal neuroblasts (cells that normally give rise to nerve tissue), and some from adult mammary glands. It was this third case with mammary cells and the fact that the cell came from an adult mammal truly captured people's attention, because it seemed to suggest that it might prove possible to clone ourselves. Of the 277 nuclei that they transferred in this experiment, only twenty-nine produced embryos; when those embryos were transferred to surrogate mothers, only one made it all the way to a live birth. This was Dolly, born in 1996. The report appeared early in 1997, in one of the most respectable scientific journals, and only after the researchers had made sure of their results and also had filed for a patent on the process.

Many commentators have reflected on why Dolly generated so much fuss. Her birth brought issues about gender: both Dolly and the donor were female, suggesting that females could do the whole reproductive process themselves.[37] She also brought many questions about the procedure's safety: it took many, many eggs and many donor cells to produce even one cloned adult. The process involved a great waste of eggs, which would be difficult to justify for human eggs as they are only obtained at real physical cost to the female donor. In addition, this case suggested the possibility that cloning might not yield a predictably good result. Presumably, the resulting clones may be healthy and normal, but they may not develop at all, or in the worst case may develop but not normally.

Immediately, bioethicists had a field day with cloning. Many jumped in with opinions, and some said foolish things, often without really understanding the science involved. There is no point in naming them here because some have since admitted that they responded before the science and its implications were clear. With such a large and lively bioethics field, not everyone can track the nuanced scientific issues that come along. Professional discretion might suggest that experts, whether in bioethics, history, policy, or any other field, should not talk on the record until they understand the issues at hand, but when reporters invited bioethicists to express their opinions on the case, many did.

The reactions reached another peak with the announcement by the South Korean researcher Hwang Woo-suk that he and his team had in fact performed transplantation research and cloned a human. This report caused a sensation at the annual meeting of the American Association for the Advancement of Science (AAAS) in February 2004, and the prestigious magazine *Science,* published by the AAAS, fast-tracked their report into publication.[38] In retrospect, even the editor of *Science,* Donald Kennedy, admitted that he and his colleagues were too eager to hear that cloning humans had happened, so their review process was less rigorously applied than normal. Such a surprising and dramatic announcement should have called for very careful questioning of the results, but reviewers can easily make assumptions based on what they already believe. At the time, many did believe that human cloning was sure to happen, and it was just a matter of when. Here they had it at last, or so it seemed.

In fact, they did not. As events unfolded over the next couple of years, it became clear that Hwang had lied about cloning a human. Investigators pointed to falsified photographs and other undocumented claims. They also discovered that Hwang's team had enticed or perhaps even coerced women associated with the laboratory into donating eggs for the research. This is universally considered unethical—it puts tremendous pressure on the women to do something risky, for which it is universally agreed they must provide informed consent.

Yet Hwang actually had done impressive cloning work, including the first clone of a dog—Snuppy—as well as having other agricultural successes with animals. Hwang also was judged to have successfully caused a human egg to develop parthenogenetically without fertilization, though he did not do this with a cloned embryo as he had claimed. After a series of accusations and investigations of his work at Seoul National University and of his publications, it became clear that he had not cloned a human embryo. Hwang claimed that his commitment to working hard and his drive to discover medical treatments had led him to overreach his evidence and to make the false claim. In retrospect, Hwang clearly knew what he was doing: along with his closest colleagues, he did actually engage in fraud.

The case evoked intense legal and ethical debate alongside the investigations of the science, of course. Hwang's case stimulated additional and more thoughtful discussion about human cloning, and it has become clear that nobody can articulate legitimate reasons in favor of taking the practical risks. In addition, cloning remains extremely expensive both in financial terms and ethical terms. Any eggs for research must be donated explicitly for that purpose with informed consent, and not many will likely donate for this purpose.[39] Prohibiting human cloning began to seem like an increasingly good idea, both in the United States and internationally. Yet even with all the debate and concerns, the international reactions remained mixed and inconsistent.

The controversy began in South Korea with the Hwang case in particular, but it spread to many countries and internationally through the United Nations. Thus, after years of debate and stimulated in part by the reactions to Hwang's claims, on March 8, 2005, the United Nations announced that after long and heated debate, the General Assembly had voted in favor of the "United Nations Declaration on Human Cloning."[40] The declaration calls for prohibiting genetic engineering techniques and "all forms of human cloning inasmuch as they are incompatible with human dignity and the protection of human life." The vote was 84 in favor, 34 opposed, with 37 abstaining or absent.

The press release explained that countries had different reasons for their votes to oppose or for abstention, in some cases because they wanted the declaration to go further but in most opposing cases because they felt that the resolution was insufficiently clear about allowing some kinds of so-called therapeutic cloning. Some countries, such as China or the United Kingdom, agreed that banning human cloning for purposes of reproduction is right and that they would continue their own ban, but they would accept cloning for research purposes and for seeking therapies. This position means working with early-stage embryos that do not have any capacity for developing into human beings because they will not be implanted into a uterus.

Although a working group had put years into adjusting the wording and rallying support for the resolution, they failed to achieve consensus on the proposal to support some kinds of cloning and not others. The resulting declaration thus leaves it up to individual countries to address what is meant by "human life."[41]

Yet a declaration by the United Nations is just that. It does not have the force of law, but merely expresses the position of the international group. Therefore, in the United States researchers can still legally clone humans, should anybody want to. But given the challenges of mastering the technology, obtaining the large number of eggs required, and persuading the leading researchers to become involved, cloning is unlikely to happen any time soon. Leading researchers do not want to invest their time and effort in a low probability venture, and especially one that nearly everybody sees as a bad idea. Since the U.N. resolution, but not primarily because of the resolution, almost nobody has argued in favor of cloning humans. Although nobody had really argued for it before, many *had* argued in favor of scientific freedom and did not want to create restrictions against even this type of research. Just because it has not happened yet and does not seem to be a priority now does not mean that it cannot be done with enough tries. It just does not seem worth the investment, since this is a possible technology with no clear market interest.

One surprising twist to the story is that for many South Koreans Hwang remains a hero. The devastating revelation that he engaged in scientific fraud as well as socially unethical behavior actually gained him supporters. There are many complex reasons for this, and they offer a useful cautionary tale about how we value science. Sometimes the public so keenly embraces particular kinds of success that we are willing to look past more serious questions, just as sometimes we become so skeptical or opposed that we overlook real possibilities. All of us love heroes, and we love to hear about medical advances and amazing discoveries. South Koreans may especially long to have scientific heroes.[42] This case reminds us to keep an open and skeptical mind, to demand evidence and ask lots of questions when something seems unlikely. Hidden assumptions can hold back—or make possible—scientific progress. If the social contract for science is going to push toward more use-inspired research, we must take care not to let our enthusiasm lead us into believing results in the absence of scientific tests.

In 2013, once again scientists announced success with cloning humans, in this case just long enough to produce stem cell lines. A team led by Shoukhrat Mitalipov at the Oregon National Primate Research Center reported in a much-heralded paper in *Cell* that they had applied the methods they had worked out for primates to develop human cloned embryos.[43] Gretchen Vogel in *Science* reported, "This time it looks like it's for real."[44] At the same time, David Cyranoski announced in *Nature* that "Mitalipov and his team have finally created patient-specific ESCs through cloning, and they are keen to prove the technology is worth pursuing."[45]

Yet stem cell researchers point out that the fuss about human cloning for "therapeutic purposes"—that is, for the production of stem cell lines—is less intense now than it was when Hwang first claimed to have succeeded. Advocates for the research are less passionate because now we have alternatives, which may even be better in some respects. Opponents are less passionate because the research was clearly performed only to produce stem cell lines and with no intention of cloning a human more fully. Nonetheless, this cloning success, assuming that it is verified and duplicated and that some of the apparent sloppiness in the original publications are resolved, shows once again just how flexible human development actually is. Cloning and the reactions to it show the potential both for engineering life and for engaging in thoughtful discussion about the social implications and appropriate reactions.

Cloning Back from Extinction

As a side note, cloning has also been welcomed as a way to clone extinct species back from extinction or endangered status. When it was released in 1993, the film *Jurassic Park* tantalized the public with the idea of dinosaurs brought back from extinction. The idea seemed intriguing, but many assumed that cloning large animals and cloning from adult cells would not be possible because of the prevailing belief that differentiation works in only one direction. But in 1997, after Dolly the sheep showed that it was possible to clone mammals and to do so from adult cells, the tantalizing dream of reproducing dinosaurs gave way to a new idea. The late 1990s brought a growing public concern about lost biodiversity along with a felt imperative to do something to recover the lost genetic information. Cloning seemed to offer a new approach to preserving biodiversity, and several groups took up research to make this possible.

In 2001, Noah the cloned gaur, an endangered wild ox native to Asia, was born but lived only a few days.[46] Researchers from Advanced Cell Technology in Massachusetts had taken a skin cell from an individual gaur and added it to a cow's cell, then gestated the embryo in a cow. The nucleus was thus from the gaur, but the rest was from the cow cell or surrogate.

Other attempts followed, with a recent success in bringing back an extinct animal type. The Lazarus Project at the University of New South Wales, Australia, reported in March 2013 that they had taken cells from a dead frog and successfully cloned those cells to produce embryos of the unusual species.[47] Though the organisms did not live long, the technology is there to make it possible. Now researchers and the public are debating whether, how, and under what conditions we should want to "de-extinct" species. Biologist and popular writer Carl Zimmer reflected on these questions for *National Geographic* in 2013, in "Bringing Them Back to Life," asking, "The revival of an extinct species is no longer a fantasy. But is it a good idea?"[48]

Stem Cells in Mice

Yet another kind of engineering based in transplantation techniques—and another case where the hypes and hopes have often outstripped careful understanding—is through stem cell biology. In this case, stem cells are transplanted outside the organism, as Harrison had done with nerve fiber cells (which were in fact a kind of stem cells) back in 1907. The cells cultured this way, using different media as "food," allow researchers both to understand the normal cycle of cell development and also to present the possibilities of using the cultured cells by engineering them to do what we want them to do. After considerable trial and error in many laboratories, researchers worked out which kinds of culture media and conditions will give rise to which kinds of cells. That is, for the pluripotent stem cells that have the potential to become any kind of cell but not all kinds, scientists have learned which conditions cause the cells to become heart muscle or nerve or pancreatic cells. This work first began in the 1980s with mice. As understanding increased about the nature of the bone marrow cells and what it meant for some of them to be hematopoietic "stem cells," studies of these transplantations began to overlap with studies of stem cells in other organs and in mice.

The term *stem cell* had already been introduced at the end of the nineteenth century, initially in plants. Stem cells meant undifferentiated cells that could self-renew and with the capacity to differentiate differently depending on the conditions. The observation that cells can differentiate in response to the environment supported Driesch's views about the considerable plasticity of embryonic cells and the ability of the whole organism to regulate differentiation of its cellular parts.

The stage was set for discovery of stem cells with even greater capacities than those of the hematopoietic cells. At the Jackson Laboratory in Bar Harbor, Maine, Leroy Stevens made that discovery while working on mouse cells and development. Stevens has largely avoided the press in recent decades, but one journalist did capture at least part of his story. Stevens had received some funding from tobacco companies and asked the question whether the tobacco itself or perhaps the cigarette papers caused the cancers that occurred in mice exposed to cigarettes.[49] In the course of this research, Stevens discovered one strain of mice, number 129, in which some of the cells behaved differently from normal (Figure 6.3). Many of the mice in that strain developed teratomas, most often in the testes. This meant that teeth, hair, and other cells not normally found in the testes appeared there, apparently having differentiated in an abnormal place and at the wrong time. Stevens bred this strain to determine which cells led to the strange results.

Stevens continued his research and felt largely ignored, though in 1970 he reported on the origin of the teratomas in work that has been cited frequently since. The cells from the blastocyst's inner cell mass in that strain, from mice that were 3 to 6 days old, remained pluripotent longer than normal. That meant that they could duplicate themselves and also could become differentiated as any (but not all) kind of cells, including those making up the teratomas. When Stevens grafted the embryos to the testes of adult mice, the result was teratomas that "resemble in every respect the spontaneous testicular teratomas characteristic of strain 129/Sv."[50]

Though Stevens felt isolated and though others did not yet understand how important these pluripotent cells were, not all other embryologists ignored him. Beatrice Mintz and Karl Illmensee visited

Figure 6.3. Leroy Stevens at the Jackson Laboratory in Bar Harbor, Maine, 1977. From the 1950s to the 1970s, Stevens studied mouse mutations, leading to his discovery and naming of pluripotent stem cells. The Jackson Laboratory, which provided this photo, wishes to note that modern researchers should wear gloves when working with mice. Used with the permission of the Jackson Laboratory.

his laboratory and borrowed some of these unusual mice to adopt and adapt Stevens's methods. When they applied the same approaches, they found "an unequivocal example in animals of a nonmutational basis for transformation to malignancy and of reversal to normalcy. The origin of this tumor from a disorganized embryo suggests that malignancies of some other, more specialized stem cells might arise comparably through tissue disorganization, leading to developmental aberrations of gene expression rather than changes in gene structure."[51] Under certain very controlled environmental circumstances, the embryonic stem cells would give rise not just to the teratomas that Stevens had found, but also to whole organisms. As journalist Ricki Lewis revealed, "Their surprise announcement of this feat at a meeting floored Stevens, a story unto itself."[52]

In 1981, with mouse studies by Martin J. Evans and Matthew H. Kaufman in Cambridge and Gail Martin at the University of California, San Francisco, the lines of research that led to human pluripotent stem cell culturing began. Evans and Kaufman concluded definitively that "We have demonstrated here that it is possible to isolate pluripotent cells directly from early embryos and that they behave in a manner equivalent to EC cells isolated from teratocarcinomas. The network of inter-relationships between the mouse embryo and pluripotent cells derived from it has previously lacked only the direct link between the embryo and cells in culture for completion. We have now demonstrated this."[53] Further, as Martin concluded, "Given these results, it seems likely that there will soon be available pluripotent, embryo-derived cell lines with specific genetic alterations that should make possible a variety of new approaches to the study of early mammalian development."[54]

The story of stem cell research reaching from mice to men is well known, so we need not repeat that story or that of the controversial reactions to the research now, except to recall that the work of James Thomson and John Gearhart demonstrated in 1998 the same ability to culture pluripotent stem cell lines in humans as well as mice.[55] The major difference between 1981 and 1998 was not the extension of the techniques to human cells, but the advances in understanding the underlying genetic basis for development. Molecular genetics had progressed to the point that in less than a decade after establishment of the first isolated human embryonic pluripotent stem cell lines, researchers had discovered ways to isolate the essential genetic factors and induce pluripotency in cells that came not just from embryos but also from adult cells. We will return to this research in the context of current stem cell research.

Testing Genetics and Development: Preimplantation Genetic Diagnosis

We have discussed examples of transplanting pieces from one embryo to another, or combining cells from two embryos to produce a new hybrid whole that develops in essentially the same way as normal embryos. Another example, first with animals and then with humans, shows the strength of the integrity of the embryo as a whole and of the resulting whole organism. Regarded as a medical innovation rather than experimentation, preimplantation genetic diagnosis (PGD) served clinical therapeutic purposes and thus has been counted as a treatment. This innovation made it possible to perform genetic diagnosis at the blastocyst stage, before an embryo is implanted into the uterus. As mentioned earlier, PGD also made possible considerable scientific understanding of early cell division and early embryos. This new procedure allowed researchers to remove one or two cells from the eight-cell stage to be tested separately. In so doing, researchers could also observe them and learn a great deal about developmental stages in addition to the genetic characteristics that were the target for the procedure.

It took some courage and considerable willingness to take a risk the first time the test was used; as we have seen, the only way to test the safety of a technique is to try using it. By now, however, thousands of babies have been born after such tests, who seem to be normal and unharmed. In some cases, the test uses two of the eight cells rather than only one, and using up to two cells has now become the accepted clinical protocol in the United States and in other centers. The remaining six or seven cells still make up a whole preimplantation embryo, which can be frozen temporarily to wait for results or implanted should the embryo prove "healthy."[56]

Implementing the PGD form of early genetic testing takes us back to in vitro fertilization (IVF) treatments for infertility. At first, the research focused on giving infertile couples an opportunity to have children of their own. With time and further research, however, it also became possible to ensure that the embryos available for implantation were healthy. Genetic testing has provided considerable room for discussion about what we mean by healthy, including what might serve the eugenic purposes of public health as well as the intentions of the future child and parents.

In considering what being healthy means, genetics and embryology come together very clearly. The previous hereditarian assumption had held that because some conditions (though only a relatively few, such as cystic fibrosis or phenylketonuria, each of which has its own qualifications and questions) seem to be caused by one defective gene, we would only need a genetic test for that particular gene to determine whether an embryo carries it (and hence the condition). By the turn of the twenty-first century, it had become clear that such a simple hereditarian view did not conform to reality. The relevant issues in this rich history include examining assumptions about what counts as defective, the efficacy and risks of genetic testing, and reproductive rights and responsibilities, as well as gender politics more generally. Ideas about constructing and engineering embryos start from a mix of views about what is actual, what is possible, and what is good.[57]

Setting aside the ethical and social-political factors, the idea of testing embryos to determine whether they carry particular genes seems as though it could yield useful information. But how can we test an embryo safely and reliably? An uncertainty problem rises here: doing the test at the fertilized one-cell stage requires obtaining genetic material from inside the cell. How could we test an embryo without taking a cell and thereby destroying the embryo with the test? We saw with Roux's experiment that when he killed one of the two cells of the frog egg only half the embryo developed, and the same with the four-cell and eight-cell stages. Although Driesch had shown that sea urchin eggs could develop into full embryos, even those came out a little smaller than normal. Thus, taking one of the cells seemed as though it would impact the remaining cells and the remaining embryo. It did not seem moral, nor did it seem wise to experiment and risk causing such damage. Yet someone did with PGD, perhaps persuaded by successful experiments with mice that human cell removal could be performed safely as well.

Testing mice showed that cells in these mammals did not behave exactly like frog or sea urchin cells. Instead, the cells remained totipotent up to the eight-cell stage, which offered a great opportunity to try taking one cell and testing it while leaving the rest of the embryo to develop—hopefully normally. Mintz's work had reinforced this possibility, suggesting that the cells up to that stage can act largely independently. In the 1980s, the development of polymerase chain reaction techniques to test for particular genes made it possible for clinics to start testing.

By the 1980s, all these different and sometimes related lines of research had generated a greatly improved understanding of embryos and how they develop, though many questions remained. Researchers could take embryos apart, put the pieces together in different ways, genetically engineer, and even test them. Innovations through the 1990s brought new technologies, including cloning, innovative stem cell research, development of organs on a chip, efforts to create cells, and other research that falls under the category of regenerative medicine. E. B. Wilson's words from 1896 remain appropriate: "I can only express my conviction that the magnitude of the problem of development, whether ontogenetic or phylogenetic, has been underestimated; and that the progress of science is retarded rather than advanced by a premature attack upon its ultimate problems. Yet the splendid achievements of cell-research in the past twenty years stand as the promise of its possibilities for the future, and we need set no limit to its advance."[58]

Human Genome Project

Another project was inspired by hopes for controlling and engineering cells, including embryonic cells, but focused deeper inside the organism—into the nucleus and chromosomes. The Human Genome Project began in the 1980s and brought enthusiastic hopes that transcribing the genetic code locked within the genome would lead to a blueprint for heredity and development. James Watson led the energetic call for sequencing the genome, the same Watson who had worked alongside Crick to discover the double helical structure of DNA. In 1988, Congress voted to fund the human genome project, and Watson was appointed its director with funding provided through the National Institutes of Health (NIH) National Center for Human Genome Research. As expected, Collins became an ardent advocate for the project, who proved himself eager to speak vehemently about the importance of the project at every opportunity. Only when changing politics brought new leadership to the NIH did Watson come into serious conflict with the new director Dr. Bernadette Healy over patents and access. Watson wanted the scientific information to be open and available; Healy argued that if somebody was going to benefit from patents, it should be the

NIH. Watson left, and the project continued under the leadership of Francis Collins, who later would become the director of the entire NIH.

Some critics have pointed to the rise of genetics and genomics with its focus on testing and describing individual genomes, and especially to the argument for studying genomics to identify and prevent "disease," as a "new eugenics." Some critics at the time complained that the project involved such "big science" that it sucked up funding that should have supported other projects that would more directly impact health outcomes for more people. When the results of two different versions of the sequencing appeared in print in 2001, appearing in both *Science* and *Nature* at the same time, some of the criticism diminished. Thoughtful reflections on the genome project and its impact a decade later removed some of the initial worries, but the availability of genetic information in the absence of clear interpretation of what it means or of useful clinical applications has raised other issues.[59]

Some new issues surround the patenting of human genes, for example, and the sheer volume of private funding that has flowed to genome sequencing and to carving up ownership for different groups and purposes. After very lively debates, the U.S. courts ruled that companies could patent gene sequences. The decisions held that it was not the biological material that was being patented but the information acquired through a process that warranted intellectual property protections. James Watson, who was then heading the Human Genome Project, disagreed with the then-director of the NIH Bernadette Healy about the patentability of the information and also about whether the NIH should support such patenting. Watson argued that the information should be open. It is thus unsurprising that Watson was again a vocal spokesman for openness when the issue returned to the U.S. courts in 2012. We are likely to see continued debate about the issue, which will eventually also affect the patentability for other kinds of embryo-related data.

Many have embraced the availability of genomic knowledge, even though they do not assume that genes define us and even when they realize that translating genetic information into clinically useful forms has proven and almost surely will continue to prove elusive. Such supporters see the information about what genes we carry as valuable background that may eventually have medical advantages. The older generation has worried about the privacy of genetic information, while the younger generations seem to have accepted the ubiquity of personal exposure and feel the increased knowledge is worth the risk. Consumers have eagerly embraced the services of such businesses as 23andMe, which produces individual genetic reports from saliva samples. The company makes the report available online and updates the information as new results become available. The company website explains its privacy policy, and conspiracy critics will no doubt never believe the reassurances, yet those eager for genetic information clearly do accept the risks:

> Your personalized 23andMe web account provides secure and easy access to your information, with multiple levels of encryption and security protocols protecting your personal information. The Genetic Information Nondiscrimination Act (GINA) is U.S. federal legislation that protects Americans from discrimination (in health insurance and employment decisions) on the basis of genetic information.[60]

One lawyer, who also holds a doctorate in genetics and who gives many lectures about genetic information and privacy, points out that when the test first came out, he would go online and click on his own results to show his law students how it works with his own information. But the company updates the information so often now that he realized that he might be demonstrating to an entire audience that he had the factor for some new disease trait that he had not known about before the lecture. So whenever news arises about new correlations of genetics and disease, he checks out the results right away. He is an enthusiastic advocate for making genetic information available to individuals, wants to know his own, sees the importance of protecting privacy, and fully realizes that such data can easily be over-interpreted by those without much training in either genetics or medicine.[61] The appearance of rich information about individual genetic information creates an environment that makes a new and different kind of eugenics possible.

Obviously, this discussion does not focus directly on embryos or development but rather on the genes and genome that help direct what happens during development. Yet as the availability of genetic information increases, and as those holding the information use it to make reproductive decisions, some embryos will be viewed as having genes that are more valuable or desirable than others. Already there are parents wanting to choose the sex of their offspring, and parents wanting to avoid certain disease-bearing genes. As we will see, genetic information can provide a background for engineering or even for creating embryos with particular specifications. As bioethicists have pointed out, we should keep this possibility in mind and think individually—and also, ideally, as a society—about what kinds of outcomes we want. We should do that thinking early and often to clarify our own values and assumptions before real cases appear that demand immediate and difficult decisions. The need for serious reflection increases as researchers move from thinking about the potential for engineering toward the ability to carry out actual controlled, created, and engineered life.

The twentieth century took us from the idea of engineering life to the actual practice of engineering cells, tissues, and controlling as many factors related to development as possible. The 1890s brought a Progressivist excitement about the possibilities for progress through science and engineering, with a particular focus on eliminating undesired traits. That energy has reemerged at various times since. In contrast, but also in the same tradition, the 1990s brought excitement about the possibilities for stem cell therapies through transplantation technologies and creative engineering solutions. The focus on studying the human genome combined aspects of both approaches. Of course, the details of heredity and development also have turned out to be more complex than researchers had hoped. Yes, many of the questions are tractable in the sense that careful biological research can bring answers and help to refine our questions so that we can move ahead to new answers. Yet each step requires many pieces to work together, and the mecca of biological engineering had not arrived yet by the 1990s. The next chapter explores the following decades, taking us to where we are today.

7

Constructing Embryos for Society, Stem Cells in Action

In the past few decades, embryos have received more media attention than during the preceding millennia. Previously imagined prospects for controlling life now have come true in many different ways that involve embryos. By the time this book appears in print, still more new discoveries and new questions undoubtedly will have emerged, some of which will confront us with what seem to be new questions or with calls for new kinds of answers. Yet those discoveries, questions, and answers all exist in an historical context. Reflecting on each new development in the light of history can inform our understanding of how current science intersects with society.

One fundamental assumption in modern times, beginning with the European Scientific Revolution and the Enlightenment periods of the seventeenth and eighteenth centuries, has been that governments have a role in funding scientific research. Indeed, according to this assumption, governments have a responsibility to invest in discovery and innovation. This governmental role has remained largely unchallenged, at least in its most basic form that assumes that some public funding should be available for research in science and technology. Yet the exact scope of such funding and which sciences and what technology should be funded have raised vigorously debated questions for a variety of reasons.

At times, notably during the politically volatile 1960s, protestors in the United States and Europe especially would argue vigorously against funding weapons research and development. Critics have often called for limits on the magnitude of federal expenditures in areas of technology and medicine, arguing that industries and private companies should invest their own funds because they will benefit. The Bayh-Dole Act in the United States, officially but less memorably named the Patent and Trademark Law Amendments Act of 1980, laid out guidelines for the process by which federal investment in innovation could lead to patents and financial benefits for individual companies. Gene patenting raised new questions, as federally funded gene sequencing led to private company patents—at least until court rulings severely restricted the reach of such patents. Thus, we see that the general agreement that federal governments should invest in science and technology has met with specific areas of challenge.

Embryo research has created challenges of a different sort. Human embryonic stem cell research offers tremendous promise, both through basic research into understanding how development occurs and through applied research into discovering therapies. Yet embryonic stem cell research requires embryos, so such research runs head on into questions about the meaning of embryos. Are embryos little persons, as the metaphysically inspired, theoretical interpretation insists? Or at their earliest embryonic stages are they just material cells in a dish? Or are they something in between? What is at stake in performing or not performing research on them? What is at stake if the government invests federal funds rather than relying on private investment in such research?

This brings us back to the question of why the government funds research. Under what conditions does society decide to invest in one area of research rather than others? What regulations lay out the boundaries and guidelines for research? This chapter begins by looking at the rationale for government investment, in particular examining what is called the social contract for science and society. It then focuses on human stem cell research, looking at the different kinds of stem cells and at the research being done in each case. This branch of science exists in a highly politicized social context, so discussion of the bioethical and legal regulatory contexts plays a central part in this story. Here, science and society are in direct contact. The chapter ends with a look at regenerative biology and the potential medical applications, which pave the way for the final chapter that looks forward.

The Social Contract

Scientists have not always busily observed and experimented in order to learn for the sake of learning. Not all science has arisen as so-called curiosity-driven or basic research. With embryo research, we saw that Jacques Loeb had a keen interest in engineering life, and clinical medical researchers since have remained eager to discover and develop therapies that work. Yet until quite recently, there was more hope than action. We simply did not know enough to construct embryos or to engineer them to do what we want. Yes, some of the discoveries with mice or in agriculture had practical applicability, but only with further work on cloning, stem cell research, regenerative medicine, and related research have real prospects appeared for major advances with constructing embryos. This raises the questions of who is funding research and why. This takes us to the social contract for science and the goals of scientific research, which have themselves undergone important changes.

After World War II, thanks largely to the efforts of engineer Vannevar Bush, who became a science administrator and policy maker, the United States established the National Science Foundation (NSF). In fact, growing support for federal funding for scientific research already had led to the establishment of the National

Institutes of Health (NIH) starting in 1930 (at first just one Institute, but later plural Institutes). This effort had involved considerable congressional and public debate about who should pay for research, along with a growing social interest in public funding.

One senator, Henry Kilgore from West Virginia, had a strong interest in seeing public funding for science. Starting in 1942 and continuing until he succeeded with a bill signed into law in 1950, Kilgore began introducing the legislation that would establish the NSF. At first, he could not attract enough support, largely because he favored spreading the funds around through grants and targeted contracts for what he considered basic and also applied research, with the work to be overseen by political leaders. Some opponents objected to federal funding of science at all, and others wanted to concentrate the funds in a few places to have more impact or to favor their own constituents. Others wanted to support only basic or only applied research, and still others worried about having the process overseen politically. Kilgore had given his opponents too many reasons to reject his proposal.

An alternative model for a social contract then emerged and ultimately prevailed. This approach involved publicly funding scientific research and letting the scientists evaluate the projects based on scientific merit alone. Peer review would prevail on this model—which persists today. To avoid waste and misguided investment, the reasoning went, political processes would not decide which projects to fund. The research would remain basic research, with applications to follow as appropriate (and as pursued eagerly by private enterprise) rather than targeting particular applications from the start. From this point of view, not enough is known about the target areas for development until more research has been done. Vannevar Bush became the spokesman for this model.

A brilliant administrator and as director of the Office of Scientific Research and Development for the White House, Bush persuaded President Franklin Delano Roosevelt to charge him with writing a report on the best way for the country to support science after the end of World War II. Bush's report, *Science—The Endless Frontier,* provided a clear, strongly argued document for a traditional form of evaluation for science that relies on review by peers.[1] A National Science Foundation, he said, should support science but not for political reasons nor on grounds of expediency. Rather, it should support only the very best "basic research in the colleges, universities, and research institutes, both in medicine and the natural sciences, adapted to supporting research on new weapons for both Services, [and] adapted to administering a program of science scholarships and fellowships."[2]

Bush called for federal investing in basic and curiosity-driven fundamental research because "the frontier of science remains. It is in keeping with the American tradition—one which has made the United States great—that new frontiers shall be made accessible for development by all American citizens."[3] Thus, support with government funds based on the merit of the proposals would allow the most talented citizens to study and then carry out the scientific research. This was an inspiring goal, and Bush succeeded in persuading political leaders and Congress to approve the plan. The NSF was authorized, then established and funded starting in 1950. The authorization language called for the NSF "to promote the progress of science; to advance the national health, prosperity, and welfare; to secure the national defense."[4]

The subsequent development of the NSF into a major research agency, which reports to the executive branch of the United States government through its director who is the only NSF political appointment, has continued to emphasize the importance of foundational, basic science as its key mission. This social contract for science says that the federal government will fund the basic research, which then will be developed into practical applications by business and private interests. This way of doing the business of scientific research has led to much discussion and negotiation over the years about ownership of the results that were supported by federal money. The separation of basic science from its applications has nonetheless served as the prevailing approach of government funders.

In 1997, the political scientist Donald Stokes stimulated much public discussion of this social contract and whether it should be reconsidered with what he called Pasteur's Quadrant, which he based on his interpretation of the ideas of French chemist Louis Pasteur. Pasteur had rejected any notion of a neat or even real distinction between basic science and use-inspired science. He had insisted that he did both at the same time himself. Yet historians have shown that Pasteur was not quite the inspiring example that commentators such as Stokes saw him to be. According to some interpretations, Pasteur engaged in scientific fraud, or at the least he exaggerated and reported results before he actually had them scientifically confirmed. For example, historian Gerald Geison explains that Pasteur was so eager to use his rabies treatment for patients that he exaggerated his reassurances to the public that he had tested it carefully and knew the treatment to be safe.[5] He did this to gain financing and public support for his own work and to be able to make the claim that research pays off, even though he had not directly proven that yet.

Stokes challenged Vannevar Bush's reasoning that government funding should support primarily basic science, instead suggesting that use-inspired research is also fundamental for the health of the scientific enterprise and for society.[6] In other words, he rejected any neat dichotomy or competition of basic versus applied science. Stokes challenged the American research community to rethink their assumptions and change the way they do business, saying it is fine and even desirable in many cases to start from problems crying out for solutions and to focus on possible applications of research, and then the development community of private business can join federal funding agencies in carrying out mission-driven and use-inspired research. He was against pretending that there is always an underlying "pure" or basic science motivation, insisting instead that use-inspired research can also yield answers to questions that will in turn add to our foundational understanding.

Some rejected Stokes's ideas outright and insisted that pursuing selected, defined uses or missions would prove dangerous. In this view, by putting federal funding eggs into the wrong baskets, we could miss out on other opportunities. In addition, it is hard to know which problems are likely to prove solvable and which uses are likely to prove genuinely useful. These opponents reasoned that individual scientific researchers themselves know best what paths will likely prove the most productive. Other opponents suggested that rocking the boat might become politically dangerous, in that the public eagerly supported scientific and medical research, and that as long as the social contract was "working," we should leave it alone. As long as the existing system provided them access to the money they wanted, they did not want to risk changing that system. Still others allied themselves with Stokes's challenge to the scientific establishment but took it even further—in their view, letting scientists run the scientific enterprise might be a mistake, the equivalent of letting the lunatics run the asylum.

From this continuing swirl of healthy discussion about the proper role of science in and for society, a few leaders have emerged. In particular, Michael Crow, the president of Arizona State University, has led the call for the academic and educational enterprise at large to keep in mind at all times the possibilities for use-inspired outcomes.[7] Crow founded the academically grounded, politically informed, and publicly active Consortium for Science, Policy, and Outcomes, which takes up the direct charge to assess the social outcomes of scientific research. Crow and his colleagues, including Daniel Sarewitz, argue that research for its own sake—which obviously makes scientists happy—is not enough, particularly when we could be engaging in research directed toward solving real-world problems and having broad impact.[8]

We will have to wait to see how this discussion will play out, but in times of limited funding, the public can reasonably expect scientists to justify public investment in their research. Indeed, presently the NSF has decided to implement revised evaluation standards to ensure that proposals include more effective articulation of the purposes for the research. Science, the NSF leaders urge, should be transformative in positive ways. It should inspire results, as one recent NSF cross-cutting interdisciplinary program called Integrated NSF Support Promoting Interdisciplinary Research and Education (INSPIRE) has urged.

Meanwhile, the National Institutes of Health (NIH) have an explicit mission "to seek fundamental knowledge about the nature and behavior of living systems and the application of that knowledge to enhance health, lengthen life, and reduce the burdens of illness and disability."[9] Under a previous NIH director, Elias A. Zerhouni, who served from 2002 to 2008, the NIH experienced a dramatic shift of emphasis toward what he called "translational research." Zerhouni, who was appointed by President George W. Bush, urged researchers to broaden their focus beyond laboratory science, insisting that the entire research system (though not necessarily every researcher) should be emphasizing products development and clinical results. "Translational" meant translating their research into use-inspired results. Otherwise, knowledge that was not use-directed would fail to carry its part of the mission to "enhance health, lengthen life, and reduce the burdens of illness and disability."

Though this shift of emphasis has brought controversy, for many of the same reasons as mentioned with the criticism of use-inspired research generally, the NIH has retained the translational focus under its subsequent director Francis Collins. In fact, in December 2010, Collins announced his plan to shift funds to a new National Center for Advancing Translational Research. A number of different individuals and groups immediately attacked his ideas for change. Most resisted any change at all, but a few called for the NIH to go further in its reforms. Again, Michael Crow led the way in urging even more radical rethinking of the social contract related to scientific research. In an editorial in *Nature,* Crow called for a radically new model.[10] Yes, translation and application of research is good, Crow wrote, but the idea of a separate institute for translation does not go far enough. Why just one institute, and why make it separate? Why not reorganize completely and integrate the goals?

Of course, Crow does not run the NIH, and he is quite aware of the political difficulties of change, much less in such dramatic ways. But he productively urges us to imagine what we might do if we could start over. Why not eliminate the division into all those different disease-oriented institutes that have arisen over time and organize into three? A Biomedical Systems Research Institute could take on big, systemic problems such as diabetes and seek both to understand them and to provide use-inspired medical solutions. Second would come the Health Outcomes Institute, which would work to bring "measurable improvements to people's health."[11] After all, why should the National Institutes of Health not look at overall health? Third would be the Health Transformation Institute, which brings together what Zerhouni and Collins have thought of as the translational component and, in Crow's words, would "develop more sustainable cost models by integrating science, technology, clinical practice, economics and demographics."[12] Together, these three agencies would support the study of living systems, disease, and more importantly also health and would translate research into clinical products that could lead to achieving desirable health outcomes.

By this point, the reader may be wondering what all this has to do with embryos. The direct connection comes through what has been called regenerative medicine, which draws on knowledge about embryonic development and also on embryos themselves, starting with stem cell research. The NIH funds much of the research into embryos and regenerative medicine in the United States. Crow's model is well designed for more effective pursuit of regenerative medicine, and regenerative medicine itself provides an excellent example of the desired goal of the social contract for scientific research.

The public wants results, ideally sooner rather than later. They are taking their eagerness to the state level in some cases. In particular, investment by the citizens of the state of California in stem cell research shows that by public vote they have willingly committed to fund research—as long as there is integration of the basic research with plans for the applied or translational results as soon as possible. Other states have also invested in stem cell research, while still others have decidedly turned away or even in a few cases prohibited that such research be performed with their public funding.[13] Such state initiatives reflect the importance of evolving social contracts for science and society, and we surely have not seen a final ideal system yet.

Embryonic Stem Cells for Society

For the public, cloning and stem cell research often appear muddled together because they initially appeared in the public media in quick succession in 1997 and 1998. In fact, some scientists made the argument that the best stem cell research would occur if it started with cloning so that the stem cell lines that resulted would match the donor for the cloning process. This argument would pin the two together, but biologically the two processes are completely different.

Cloning, as we have seen, involves adapting the techniques that Briggs, King, and Gurdon had developed decades earlier. It involves transplanting a piece from one organism to another to initiate development. In the earlier studies on frogs, Briggs, King, and Gurdon transplanted nuclei in particular from various ages of eggs and embryos and eventually from older frog stages. In 1998, Ian Wilmut and his team cloned the sheep Dolly by transplanting an entire cell from the adult mammal to an egg. After that point, once the new nucleus (and the cellular material that came with it) has been absorbed and combined with the egg host, the regulatory processes that Hans Driesch had seen a century before with his sea urchins (made from half of a two-cell embryo, as discussed earlier) begin acting to make the organism whole. The ensuing stages of development pursue the same developmental path as a normal embryo, as seen in those series of normal stages such as the Carnegie stages studied from the beginning of the twentieth century.

In cloning, development happens normally after the nucleus is replaced either with a transplanted nucleus or with an entire cell containing a nucleus. Stem cell research happens quite differently. In 1998, James Thomson announced that he had developed a culture of stem cells from human embryos, and John Gearhart announced that he had derived a line of embryonic germ cells from human fetuses. Their stem cell research was reported against the backdrop of cloning from just the year before, so the public heard the two types of research as closely related, if not the same.[14] Until then, most members of the public had not given much attention or thought to embryos.

Stem cell research refers to research using several different kinds of cells and very different research methods, including what have been labeled as embryonic, adult, and induced pluripotent stem cell technology. To be wise citizens when confronted with political decisions concerning embryos, we must understand the differences among the different kinds of stem cell research, why they are called that, and what is at issue and at stake in each case.

There are three major kinds of stem cells: embryonic, adult, and induced pluripotent. In brief, embryonic stem cells come from embryos at the blastocyst stage, and they are pluripotent; researchers produce these pluripotent embryonic stem cells by killing the egg. The second type, adult stem cells, come from any stage after the embryo, including from fetuses, and they are called "adult" because they are already on the way to becoming differentiated as a particular kind of cell and are no longer pluripotent. Adult stem cells do not have the capacity to become any type of cell; rather, they can become only cells of one or of just several types. Third are the induced pluripotent stem cells (iPSC), which begin as already differentiated cells that are then engineered in the laboratory to become genetically reprogrammed. This process causes the cells to de-differentiate and revert to their earliest state of undifferentiated pluripotency. The NIH website does an excellent job of explaining these differences, but let us look at the history of these distinctions a bit further.

When an embryo begins to develop, the first cells up to and through the eight-cell stage remain *totipotent,* each one capable of becoming the whole organism. For example, they can make octuplets, even without laboratory help, most often when fertility drugs cause the cells to separate. Obviously, octuplets do not occur very often, and the only recorded cases in humans have required technological intervention. However, as Beatrice Mintz has shown, if we separate the cells at the eight-cell stage in mice, each can develop separately, or any combination of them can develop. They can, Mintz demonstrated, even combine with other cells from other individuals.[15] And as Nicole Le Douarin showed, even cells from different species can combine together and give rise to one integrated chimeric organism.[16] Those powerful processes of organic regulation of the individual, which Driesch had noted, guide the resulting biological development. So each of the initial eight cells is totipotent, but they are also highly plastic in their ability to respond to changing external conditions.

What about after the eight-cell stage? We said that the next stage starts the cells dividing at different rates, and they proceed through many divisions up to the blastocyst stage, with its inner cell mass of undifferentiated pluripotent embryonic stem cells and a surrounding layer of cells that will make up the placenta. The blastocyst in humans, as observed in vitro, typically occurs by day five; the cell numbers continue to multiply to day nine, and day fourteen is the latest time at which a blastocyst remains able to live on its own without being implanted into a uterus. The number of cells increases during this time up to around a hundred (though the number of cells varies, depending on the point in the blastocyst's development and the external conditions, all of which shows just how flexible and plastic the developing embryo is). These are called "human embryonic pluripotent stem cells" because they come from humans, come from embryos, have the potential to become any kind of cell (and are thus pluripotent), have not yet differentiated, and retain the capacity to self-renew or duplicate.

These pluripotent stem cells thus have great capacity to differentiate, at which point they lose their pluripotency and become multipotent or unipotent (that is, they have the ability to become differentiated as either one of several or only one kind of cell). Researchers have found it exciting to experiment with these pluripotent cells and their capacity to become any kind of cell, to discover just what these cells can do. Stem cell biologists have studied the culture media and environmental conditions that cause pluripotent cells to develop into particular kinds of cells. What does it take to make them become, say, pancreatic islet cells that can produce insulin? Or become heart muscle cells? Or become neural cells? Until quite recently, developmental biologists still assumed that once the cells had been differentiated, they would remain differentiated, that development only works in one direction.

Once Thomson and Gearhart cultured their cell lines, researchers began experimenting with them and generating more such cell lines. Not only could researchers do research, but they might soon succeed with use-inspired work to generate clinical therapies. Nonetheless, biological challenges remained, such as the fact that the embryonic stem cells in the culture dish are not the same as such cells in the body because they live in a different environment. As the stem cell biologist Jonathan Slack convincingly explains in his little book *Stem Cells: A Very Short Introduction,* the kind of highly pluripotent human embryonic stem cells that have been created in the laboratory simply do not exist in the embryo.[17] As soon as these cells are extracted from the blastocyst and start to divide, they already have become different. This artificial construction of new cells creates challenges in interpreting normal processes, but it also creates opportunities. The more research they do, the more researchers will know and understand the relationships between the cells that develop in the laboratory and the cells that develop in the body.

Another biological challenge has occurred because of cell culture technology. In the initial research, when the cells were removed from the blastocyst and placed into the culture dish, they began to divide, but they were not in the dishes alone. They were thriving there because of feeder cells that provided nutrients for them. All cell cultures depend on some culture medium to survive. In this case, all the early cells lines were growing on a single layer of cells that were sticky and provided a substrate for the culture. In the case of the early experiments with human embryonic stem cells, the feeder cells came from mice. That raised questions about the safety of the procedure as well as about how close to "normal" the experimental culture conditions were. Here was another challenge, but one that was soon addressed by creating other types of culture media that did not include nonhuman cells.

Even as researchers have confronted the biological challenges, one really big problem has remained, namely, a major ethical problem. The only way to study the human embryonic stem cells, and the only way to use these cells to develop therapies, required destroying the embryos. No other process can extract the desirable pluripotent cells out of the blastocysts than to destroy the embryo. Even if somebody figures out a way to remove just one or a few cells, this experimental procedure would likely be considered unethical because without a lot of risk-taking nobody knows whether the remaining cells could develop into a healthy adult. If they could develop but not healthfully or normally, then it would be considered unethical to cause an unhealthy human on purpose. And if the researchers did not let the experimental embryos with missing cells continue to develop further, then the embryo would still have died.

Therefore, stem cell research requires using embryos, and in the process necessarily killing them.

Fertility clinics raise the same ethical dilemma because they discard, or in other words kill, embryos every day. Fertility clinics fertilize far more embryos than the parents want or can implant and gestate. The fertility business—and our tacit social acceptance of it—tolerates this loss because it allows more successful results. Early debates about stem cell research brought disputes about just how many of these embryos do get discarded, but clearly there are at least tens of thousands. The embryos exist already, and some will be discarded, so it seemed at least to some (including those whose eggs and sperm had led to the production of some of those embryos) that researchers should be allowed to use these embryos even if it means destroying them.

Somehow, though, against the background of cloning and the background of heated abortion politics, stem cell research has provoked confusion, divided opinion, and has made a lot of people queasy. Many people are uncertain what to think. Does supporting stem cell research make cloning more likely? No. Does it mean that embryos in clinics would be acquired for research without their owners' permission? No. And what is an embryo anyway? Why not use embryos for research?

We need not rehearse the full range of arguments for and against stem cell research here because they have been widely discussed in many other places.[18] For example, see essays in favor and against in *Contemporary Debates in Bioethics.*[19] The emphasis here remains focused on the biological research. What have researchers learned from human embryonic stem cell research? The answer is, a lot. Because the embryonic cells have not differentiated but rather become differentiated through the research process, scientists can observe what conditions cause what outcomes. The results remain very predictable and regular: particular sets of conditions do, in fact, yield particular results. This research by itself tells us a great deal about the details of which surrounding environmental conditions cause which specific kinds of differentiation. The research is helping uncover the ways that inherited genes combine with environmental factors to yield development.

Yet this research does not occur in isolation, and it may be combined with other techniques to address other questions. With this, we move farther into the arena of what once seemed science fiction but has become real. These stem cells can be derived from lines of cells that are genetically altered, either by "knocking out" particular genes or by using recombinant techniques to add genes. Meetings involving stem cell science involve a great deal of discussion about genetics as well as culture media and environmental conditions. Embryonic stem cell research has already yielded a wealth of information about how development and differentiation work as well as how the underlying gene regulatory systems work. The experimental methods make it possible to perform a great deal of research that has added to our understanding of the biology of embryos.

Perhaps what embryonic stem cell research has done best so far is to give us enough scientific knowledge to point us in other directions, such as showing how other sources of cells can yield therapies in ways that we would not have understood without having studied embryonic stem cells. This has certainly been true with experiments on adult stem cells and on induced pluripotent stem cells (iPS or iPSC, discussed in more detail later). Still, enthusiastic support for stem cell research hinges on hope for its potential usefulness. Lobbyists have energetically argued that great applications will arise from stem cell research, and they have attempted to justify both allowing it and funding it in those terms.

Along with efforts in other countries, in the United States the companies Geron and Advanced Cell Technology (ACT) have been the first approved to conduct a total of three clinical trials based on human embryonic stem cell technologies. For the most up-to-date information about such trials and the context of stem cell research generally, see the clinical trials section of the NIH website.[20] Both Geron and ACT are based in California, though ACT also has a Massachusetts branch and began there. In part because of the state's voter initiative to invest in stem cell research and development, California has become a welcoming place for these activities.

Performing clinical trials for drugs requires following a set of regulations overseen by the Food and Drug Administration (FDA). Stem cells have brought considerable debate about whether they are drugs and which sets of rules and regulations should hold, as well as which regulatory body should decide. The FDA asserted that it had jurisdiction; when it did, many people gave a huge sigh of relief that researchers could get on with their work with a defined process to oversee it. Geron announced their trial but then pulled back initially to address more questions. They began officially in late 2010 and began enrolling patients in October. Phase I trials test safety, and they tested in particular the safety of injecting human embryonic stem cells directly into the injured site in cases of spinal cord injury. The patients in the Geron trial must have been diagnosed as having complete spinal cord failure and must have been recently injured. The first patient received treatment in October 2010, and the second in May 2011. The preliminary reports showed that the treatment had not caused any damage, but it was too early to determine the treatment's overall outcome. Geron clearly hoped that the stem cells would travel to the injured area and help regenerate cells, thereby restoring cellular function. They ended the trial on November 14, 2011.

Developmental biologists had already focused on the retina as a likely site for stem cell therapies to work, both because retinal cells have less tendency to reject transplanted cells and because the mechanism is relatively straightforward. In addition, the eye is relatively isolated, so adding cells seems less likely to have broader systemic and possibly harmful effects. In November 2010, ACT received permission to begin enrolling patients in two separate trials designed to help cure blindness caused by macular degeneration (degeneration of the macula, or central layer of the eye). One trial focuses on patients with a relatively rare disease called Stargardt's macular dystrophy, and the other treats the much more common condition of macular degeneration caused by age. These trials began with patients in January 2011. Again, it is far too early to assess the results or whether this research will lead to other trials; however, so far the procedure seems to have passed the initial safety testing phase. The report in early 2012 was very promising and looks forward to research with even earlier stages of macular degenerative disease in hopes of rescuing cells before they are too far damaged.[21]

Obviously, just as with the debates over basic and use-inspired research, we must have patience in waiting for results from these and other trials in various stages of contemplation, planning, or early implementation. And these are just the phase I trials that test safety and how well the small number of patients can tolerate the treatment. Phase II trials typically involve larger numbers of patients, and phase III trials enroll still more, usually in formally conducted randomized, controlled study designs. Controlled designs involve giving some patients the experimental treatment and others not, then comparing the results. In specialized situations where there are few available patients to test, the FDA may allow accommodations. Yet, as many have noted, the FDA and other government agencies have not mapped out a clear set of regulatory guidelines for testing proposed stem cell therapies.

We shall see both how the trials develop and how the results turn out. Those seeking quick applications should remember how stem cells work in the body. These cells are valued because they are not yet differentiated and can be caused to differentiate in different ways, as researchers control the cells' food and environment—a process that takes time. And biologists have typically assumed that once the cells have differentiated to become particular kinds of cells, they will remain those kinds of cells—yet research has shown that they do not. They can retain their flexibility, can become de-differentiated, and can redifferentiate in new ways. Based on the scientific findings, we should not expect them to become a particular kind of cell in the laboratory dish then always remain that kind of cell after they are placed into a different environment. Even though the time-honored assumption held that cells can develop in only one direction, we now know this is not true. Under the right conditions, they can become unspecialized even after they have differentiated.

Sometimes stem cells can cause pathological conditions and serious medical problems, and these cases sound a warning to those eager to rush into stem cell therapies. As Leroy Stevens showed with his special line of mice, pluripotent stem cells that develop in the wrong place and time can lead to teratomas.[22] A May 2013 installment of a popular column in the *New York Times Sunday Magazine* described one such case: a woman who had terrible headaches and symptoms that resembled seizures or strokes, who did not improve significantly with treatment.[23] Her episodes recurred, and eventually, after a great deal of suffering for her and much frustration for the doctors, a neurologist found reports of a small number of similar cases that had involved ovarian teratoma. She had a cyst, in which a tiny tumor had grown that included pluripotent stem cells. The neurologist, and others who had seen similar cases, hypothesized that the tumor had grown neural cells, which effectively acted as brain cells and caused tremendous pain. This example should remind us that stem cells, and especially pluripotent stem cells, retain complex capacities. Any therapeutic use clearly must control misdirected developmental processes.

Embryonic stem cells may be tremendously useful as a research tool, but their clinical applications will likely not work exactly the way we imagine. Further studies with other kinds of stem cells reinforce this point and remind us of the value of conducting more research with a wider range of cells and always with an attitude of skepticism so as not to overinterpret the results. Retaining a critical eye is part of good science, and is good for developing sound applications of science for society.

Other Kinds of Stem Cells for Society

The second type of stem cell is the so-called adult stem cell. These cells include hematopoietic stem cells.[24] For decades, clinicians have used these special kinds of blood-producing cells extracted from bone marrow to treat patients with serious blood disorders like leukemias or sickness caused by excessive radiation exposure. Recall that adult stem cells are not quite what they sound like: stem cells are considered "adult" at any stage later than the embryonic stages. Even as other cells differentiate, some stem cells can persist in their undifferentiated state as "reserves" that can kick into action in case of injury or as replacements when cells age and die. That means that this kind of stem cell taken from fetuses or from umbilical cord blood before they are fully differentiated as specialized cells are considered as much adult stem cells as those taken from actual adults.

To be clear, any cell that has the two characteristics of being not yet differentiated and being capable of self-renewal (the definition for stem cell-ness), and that also appears in the context of other differentiated cells is considered an adult stem cell. Ironically, those critics of human embryonic stem cell research who oppose the study of embryos for ethical reasons often remain ignorant of the fact that when they lobby for what they believe is ethically safe research with adult cells, they may actually be lobbying for more work on cells from aborted fetuses. Of course, responsible researchers remain careful to indicate the source of cells.

Researchers have found it much more difficult to locate adult stem cells in the body than to isolate embryonic stem cells from the embryo. The embryo has that inner mass of cells all nicely packed together, but the relatively few adult stem cells appear scattered throughout and mixed with literally trillions of other cells. Perhaps scientists will develop tools to identify and extract just the stem cells, but they cannot do so today. Umbilical cord blood is packed with undifferentiated stem cells, and so are parts of the fetus as well as some areas of adults such as the bone marrow.

Given that research has proceeded for decades with bone marrow cells, we have some proven and known therapies that use them. However, the claim sometimes made about the great number of proven adult stem cell therapies is at best exaggerated and at worst fraudulent. Yes, it is true that clinicians can use adult stem cells to treat a few kinds of diseases and have some proven successes (see the NIH website for the latest report on successes as well as ongoing clinical trials), but we do not know how much potential really exists to harvest, to culture, or to use adult stem cells in medical therapies.

Various sorts of off-label use also occur, as when using bone marrow cells for conditions for which they have not yet been officially tested. Some of this use is understandable, performed by licensed clinicians for legitimate reasons that have solid scientific foundations. But we also see unfortunate examples of wishful thinking, fraud, and even medical tourism as desperate patients cling to any hope for a serious disease.

These examples pay out sometimes in regulated environments and sometimes in highly publicized ways, as in the case of Rick Perry. As governor of Texas, self-proclaimed Christian, and U.S. presidential candidate during the 2012 election primaries, Perry strongly advocated adult stem cell research. During spinal surgery in 2011, he had received an unapproved injection of his own stem cells that had been extracted from his own fat tissue and then grown in a laboratory by Dr. Stanley Jones. His experience raises questions about appropriate jurisdiction and about who holds the authority and responsibility to regulate such research, because it lies outside of what the FDA usually regulates and requires highly experimental procedures for which no clinical trials have occurred. Surgery has always allowed an experimental approach; the surgical profession assumes that each case will be a little different, so standard treatments may need modification in response to precise conditions that emerge. This reasoning argues for allowing the injection of stem cells if there is reason to think they might help. But the FDA regulates drugs and requires clinical trials to test other kinds of treatments; if stem cell therapies are like drugs, they require such trials. So far, stem cells have fallen into both arenas, and we will surely see court challenges in the coming years.

The evidence in favor of the claim that injecting one's own stem cells will help in cases like Perry's is primarily anecdotal. That is, the data consist of other patients and other physicians relating their experiences. What makes his particular case newsworthy is his advocacy and visible support for such treatments from his position in a political leadership role, as well as the fact that he has endorsed establishing a state stem cell bank and has advocated state support for a company providing clinical treatments. The state of Texas has, in fact, endorsed and supported adult stem cell research and treatments. Perry's case, involving the company CellTex, reveals the tangle of ethical, political, practical, and medical issues in such cases.

Medical tourism raises even worse tangles. Clinics in other countries lure "tourists" into treatments that they cannot get at home under any conditions, or that they can get more easily or more cheaply elsewhere. We do not yet know what stem cell treatments work for what purposes, what the limitations or side effects may be, or what damage they may cause—including possible later development of cancer. So far, only a few cases have demonstrated any definite benefit; we see many hopeful experiments with no evidence of success, and we find growing evidence of cases where considerable harm has occurred through unregulated and unproven treatments.

Another innovation is much better supported: considerable interest surrounds a new kind of experimentally constructed stem cell using genetic engineering and leading to induced pluripotency. This kind of stem cell appears to offer not only an ethically pure alternative to embryonic stem cells but also a highly engineered and controlled product using genetic manipulation. Such genetic manipulations, once considered problematic (as discussed earlier), now offer methods of making any number of different kinds of cells into stem cells. Not long ago, critics denounced the use of recombinant DNA as "playing God," yet now many people embrace this experimental approach to controlling cells and their destiny as far preferable to using the embryonic stem cells that occur naturally in the inner cell mass of embryos that must be destroyed to get them.

To many, genetic manipulation is now a preferable alternative to destroying embryos to generate cell lines, even when those embryos were going to be discarded anyway. Embryo research invites complex societal reactions as individuals bring their divergent ideas of the meaning of life to the table. The resulting mix of changing and inconsistent political positions seems a bit mystifying at times. Perhaps we should not be surprised, because embryos do evoke a mix of responses and in the United States we have not yet negotiated a shared social understanding of what embryos are and how we value them.

Excellent textbooks have appeared in recent years with up-to-date information on all these aspects of stem cell science that use the various kinds of cells described here.[25] These works lay out clearly the practices and prospects for embryonic stem cell, adult stem cell, and induced pluripotent cell (iPSC) research. As a reminder, because this is such an important innovation to understand, iPSC are stem cells engineered from what started out as ordinary cells of apparently almost any kind. They undergo genetic reprogramming with added genes, and these changed conditions cause them to dedifferentiate and regain the pluripotency of embryonic stem cells. Because pluripotency is assessed by assaying for the ability of the cells to replicate and also for the cells to remain undifferentiated, the research depends on slightly engineering the cells and then testing whether they meet those two requirements.

Often processes that seem the most promising ahead of time turn out not to work, or not to work the way the researchers imagined in the first place. So, trying something new is the next step once again. As we have seen with many of the successful researchers over time, they had to be very dedicated to trying again and again to get things to work. Now that we know how to create iPSC, the process seems relatively straightforward to carry out—though of course it did not seem that way for those who made the process work the first time: James Thomson and Shinya Yamanaka, who conducted two independent lines of research.

Yamanaka showed that he could take ordinary developed cells and by genetic reprogramming cause them to become de-differentiated and to behave as pluripotent cells. He did this first in mice, and the announcement surprised many who had not believed it was possible. As with cloning, many researchers had made assumptions about the ways that cells become differentiated, so they did not think that it was possible to induce them to develop backward, to return to an undifferentiated state. Yamanaka and Kuzatoshi Takahashi announced in 2006 that, by working with mice, they had taken embryonic cells and adult fibroblast cells and added four genetic factors, Oct 3/4, Sox 2, c-Myc, and Klf4, and then cultured them as they would for embryonic stem cells.[26] They had expected to need an additional gene called Nanog (in the geneticists' tradition of giving colorful names to genes) because it shows up in early developmental stages in key ways, but it turned out not to be required. They tested the resulting cells to determine that they would act like pluripotent embryonic cells. Yamanaka and Takahashi also transplanted the cells to mice and caused those cells to develop with all three germ layers in them, which is accepted as a test for pluripotency because the ability to generate all the germ layers is a characteristic of pluripotent cells, as Stevens had shown.

Immediately, anticipation mounted. Could somebody do this for human cells also? The answer, as it turned out, was yes. The proof arrived the next year from Yamanaka and also from James Thomson in Wisconsin.[27] This began a period of intense interest in discovering just how many and which combinations of genes have to work together to cause the reprogramming, how to introduce the genes without using viruses to carry them (as is the typical approach), and how to avoid other possibly risky parts of the procedure. Yamanaka and Thomson have remained leaders in the field, as has George Daley at Harvard. Daley has worked, for example, to create cell lines that other researchers can use, to speed the ability to compare results, and to develop clinical applications along with the gathering of scientific knowledge.

Others are pursuing various clinical trials with iPSC, attempting to answer questions about the safety of these engineered cells. For example, Rudolf Jaenisch's laboratory at the Whitehead Institute in Cambridge, Massachusetts, is developing tools for reprogramming stem cells in mice to allow multiple gene changes at the same time by editing the genome in several ways at once. This is tremendously promising as a research tool and as an approach for testing for disorders, diseases, and possible treatments because almost all diseases involve more than one gene.[28] Thanks to the NIH website, which brings together information from most laboratories, you can look up how many other experiments or clinical trials are under way now, and gauge the progress of each.[29] Because these cells are engineered in a way that seems to retain echoes of their past differentiated history, the question arises of whether they will remain true to their newly programmed state or revert at inopportune times. We will only know after a great deal more experimentation.

This point cannot be overemphasized: the leading researchers all acknowledge that their work is a continuation of those earlier studies of embryonic stem cells, which provided the foundational knowledge about how cells differentiate under different defined conditions. It is only because of knowledge gained from researching embryonic stem cells that we understand what pluripotency means for these cells or have had the idea of artificially constructing these cells. It is absolutely not the case that we could stop all study of embryonic stem cells right now and still learn everything we need to know or develop all possible applications. Continuing the work with embryos is important for the sake of gaining scientific knowledge through comparative exploration. This means that the development of iPSC technology does not replace human embryonic stem cell research. However, what may well become possible and we can hope for is that iPSC may replace the need for embryos to generate the stem cell lines for therapies. That is, by comparing embryonic cells and iPSC and studying them alongside each other, the iPSC might prove useful for clinical applications.

Who Decides What Is Best for Society?

Which stem cell approach is best? All of them, as it turns out, because they complement each other. Pursuing different lines of research and comparing the results offer the most information for science and for developing therapies. Those who argue that we should stop embryonic stem cell research because adult stem cell or iPSC research is the better alternative are not seeing the larger research context: each type of research allows the other types to proceed in the most informed possible way. The interconnection of the different kinds of research reached public attention in the reporting of a recent court case concerning what kind of stem cell research is best. In particular, district judge Royce Lamberth made an important ruling, which was then reversed and which had important implications for public perceptions of stem cell research.

First, before revealing the punch line for the court case, let us discuss the background. The United Kingdom, Australia, and a few other countries have led the way in articulating regulatory policies that capture what their societies value. The European Human Embryonic Stem Cell Registry resulted from a European effort to collaborate across member countries and to support research.[30] In contrast, what exists as U.S. policy on stem cell research comes from a series of policies and Executive Orders issued by three different presidents who have set the regulatory climate for managing stem cell research. As of 2013, Congress has passed no legislation directly related to stem cell research because President George W. Bush vetoed the two bills that had passed in 2006 and 2007.

Back when the first human embryonic stem cell lines appeared in 1998, Bill Clinton was serving as president. He called on the National Bioethics Advisory Committee (NBAC) that he had established for advice; they endorsed proceeding with stem cell research on human embryonic lines. However, interpreting what the 1996 Dickey-Wicker Amendment (mentioned in Chapter 1, passed in the Clinton years) meant for stem cell research proved tricky. Remember that ever since the amendment was first introduced, every bill that includes NIH funding has included language prohibiting the use of federal funds to support embryo research.[31] The sponsors clearly intended the rider to prohibit federal funding of research on embryos while not interfering with the big and profitable business of in vitro fertilization.

Yet what did it really prohibit? If the research was performed on embryos that had been fertilized or on cell lines generated from privately funded clinics, would further research using those embryos or cell lines be acceptable using federal funding? After all, research that did not use federal funding could create, destroy, discard, or knowingly subject embryos to injury or death. What were the limitations to the use of federal funding? The situation continued, undecided and open to interpretation, while presidents stepped in one after another to fill the gap.

Clinton decided that research with human embryonic stem cells is not, in fact, prohibited by Dickey-Wicker and is therefore allowed. As long as the production of stem cell lines did not use federal funds, no problem would arise under current terms. Harold Varmus served as director of NIH at the time, and he sought a legal opinion on the matter from the Department of Health and Human Services general counsel Harriet Rabb. In 1999, Rabb concluded that stem cells do not "meet the statutory definition of an embryo" and therefore did not violate Dickey-Wicker.[32] Additional advisory committees established guidelines, including the NIH's 2000 "Guidelines for Research Using Human Pluripotent Stem Cells."[33]

During his presidential campaign, George W. Bush had promised to change the situation—he made clear that he did not approve of destroying embryos under any conditions. Once in office, he announced on August 9, 2001, a prohibition on using federal funds for stem cell research—except on lines that had already been derived before that date.[34] In his announcement speech, Bush established his President's Council on Bioethics, with Leon Kass as chair, and asked the group to look closely at stem cell research in particular. Years later, in 2007 after he heard about discoveries using iPSC technology, Bush issued an Executive Order calling for a move toward alternative sources.[35]

Then another new president arrived, and on March 9, 2009, President Barack Obama shifted policy again with his Executive Order. He worked at "Removing Barriers to Responsible Scientific Research Involving Human Stem Cells" and asked the NIH to establish clear ethical research guidelines.[36] They did, and the research went forward.

Then, on August 23, 2010, federal district court judge Royce Lamberth made a ruling that shook the entire stem cell research community. He based his decision on some flawed assumptions worth examining. Lamberth ruled in the case of *Dr. James L. Sherley et al. v. Kathleen Sebelius et al,* that the plaintiffs met conditions for a preliminary injunction against the NIH and the federal funding of human embryonic stem cell research.[37] In other words, it looked like research on human embryonic stem cells was about to lose federal support.

Up to that point, an earlier ruling had established that researchers James Sherley and Theresa Deisher, who do work only on adult stem cells, had legal standing for their suit against Kathleen Sebelius in her role as Secretary of Health and Human Services, which oversees the NIH.[38] The suit called "for declaratory and injunctive relief to prevent defendants' Guidelines for Human Stem Cell Research ('Guideline') from taking effect." Sherley and Deisher (with Sherley then at MIT and Deisher at her own company, AVM Bioetchnology, in Seattle) were involved with some research on adult stem cells, and they claimed that they would suffer harm if the NIH directed funds to embryonic stem cell research because that would create competition for the limited funds for stem cell research to which, they argued, they had a legitimate claim. Furthermore, they maintained that both adult stem cell research and iPSC research are better both scientifically and ethically. With his 2010 decision, Judge Lamberth ruled in their favor.

This decision shocked the research community, in part because it required that the NIH immediately halt funding for human embryonic stem cell research, even for laboratories already working with researchers hired and dependent on the funding. Most importantly, the critics disputed the claim that other forms of stem cell research were better. Fortunately for human embryonic stem cell researchers, the injunction held only briefly because an Appeals Court accepted the case and stopped the injunction. On April 29, 2011, Judge Douglas Ginsburg presented the majority decision from that appeal. He wrote, "We conclude the plaintiffs are unlikely to prevail because Dickey-Wicker is ambiguous and the NIH seems reasonably to have concluded that, although Dickey-Wicker bars funding for the destructive act of deriving an hESC from an embryo, it does not prohibit funding a research project in which an hESC will be used."[39]

As a result, the case went back to Judge Lamberth. On July 27, 2011, the judge dismissed the case and concluded that "This Court, following the D.C. Circuit's reasoning and conclusions, must find that defendants reasonably interpreted the Dickey-Wicker Amendment to permit funding for human embryonic stem cell research because such research is not 'research in which a human embryo or embryos are destroyed.' "[40] Yes, the embryos are destroyed, but if they have been destroyed using private funding, then research on the resulting cell lines does not involve actual destruction of embryos using public funding and therefore does not violate Dickey-Wicker. This ruling helped tremendously in finally clarifying that Dickey-Wicker does not apply to stem cell lines that were derived without use of federal funds.

That is, the ruling upheld the government's claim that research can be separated into two distinct parts. The first involves using embryos and culturing stem cell lines, and the second involves performing the research with those lines. Once the embryo has been destroyed to produce the stem cells, it is no longer an embryo by this reasoning. So long as the stem cell production occurs in laboratories without any federal funding, the work does not violate Dickey-Wicker. Research on the stem cell lines themselves can use federal funding because it does not involve research on embryos as such.

An even more striking feature about the case, however, was that this federal court judge felt himself placed in the position to make such a ruling on a matter of science and that he was willing to do it. He assumed the role initially to rule in favor of a case that made the claim that iPSC and particularly adult stem cell research is "better" than embryonic stem cell research. His assumption created a very dangerous situation that flew in the face of the long-standing (though evolving) social contract for science. Society has understood since at least the Scientific Revolution that science has a set of rules about what makes some science better than other science. Governments and the public have agreed that the scientists who know those rules should police the research community through rigorous peer review. U.S. scientific efforts would suffer a serious blow if we were to allow the courts to make decisions about what is good science. Fortunately, the final ruling upheld the social contract for science that allows peer review among scientists to use the best available scientific evidence to determine what is good science.

Regenerative Medicine for Society

Stimulated by the rise of stem cell research, the NIH launched an initiative for regenerative medicine. The hope is that lost or missing function in patients with injuries or diseases can be restored, and the effort focuses specifically on using stem cell technologies. For example, in cases of injury or disease-related loss of function in the nervous system, heart muscles, or pancreatic islet cells that produce insulin, it would be great if cells could be administered to regenerate the lost function. Clearly, this exemplifies use-inspired research, and from the start it has looked very promising.

However, from the beginning of the idea, questions have persisted about how best to perform such research so as to lead to clinical applications, and how far to go with regeneration. What, exactly, is to be regenerated? In 2006, the NIH posted a report, "Regenerative Medicine," to address such questions.[41] It includes chapters by different experts and addresses the central question of what stem cells are; it then looks at specific examples: bone marrow, nervous system, genetically modified stem cells, intellectual property, cardiac disease, and diabetes. The editors later added chapters to address new issues, such as alternative approaches (iPSC), the role of stem cells in cancer, and the promise for iPSC in leading to treatments.

The chapters all do an excellent job of laying out the issues and the realistic prospects in very clear and direct terms. The authors do not exaggerate the clinical promise of stem cell research though they clearly retain considerable hope. The cover is most striking because it consists almost entirely of a compelling picture of Prometheus (Figure 7.1). We recall Prometheus from Greek mythology as a Titan who befriended mankind. The gods forbade Prometheus

Figure 7.1. Cover image of the National Institutes of Health's report on regenerative medicine featuring Prometheus. The NIH produced this report on the promise and realities of regenerative medicine using stem cell science. The first edition appeared in 2006, and the NIH has added chapters since to keep up with this rapidly expanding field. The report is publicly available at http://stemcells.nih.gov/info/scireport/pages/2006report.aspx. Image used by permission of the artist, Terese Winslow.

to share the secret of fire with humans, but he defied the prohibition and taught humans the secret. As punishment, the gods chained Prometheus to a rock, and every day an eagle would come to peck out his liver—the NIH cover illustration shows the eagle at work—and every night his liver would grow back, thus illustrating the power of regeneration.

But what, precisely, is the intended message? That we too can have our livers pecked out and regenerate new ones every night? That doesn't seem all that attractive—would there be any way out of such a dilemma? Or did the editors intend the message that livers have regenerative capacities and that we should work to discover how to harness that ability in other organ systems? The dramatic choice of image shows effectively (and probably unintentionally) that regenerative medicine brings as many questions as answers, and that not all innovations are unambiguously good.

What impact does thinking in terms of regenerative medicine have? The impact appears to be highly significant, actually. Typing the term "regenerative medicine" into Google in 2011 yielded almost 3 million hits. A year later, the number had risen to nearly 8 million, coming from the many new institutes, programs, graduate programs, and funding opportunities in this field. What, then, is the field? Mostly it pertains to stem cell research and its applications. It points to a huge research enterprise dedicated to studying the biology of stem cells that also focuses on developing therapies and applications. This growth goes along with the NIH shift to emphasizing translational research, and we see that stem cells have pushed this science in the direction of use-inspired research with the potential for clinical applications.

Table 7.1 illustrates the shift in funding, where ARRA refers to the American Recovery and Reinvestment Act of 2009, which was designed to infuse funds into the economy and promote recovery from a major recession. In 2002, the NIH provided little support for human embryonic study, in part because the Bush administration permitted researchers to carry out federally funded research

Table 7.1. Shift in National Institutes of Health funding, FY 2002–2010 (in millions of dollars).

Year	Human stem cells		Nonhuman stem cells	
	Embryonic	Nonembryonic	Embryonic	Nonembryonic
2002	10.1	170.9	71.5	134.1
2003	20.3	190.7	113.5	192.1
2004	24.3	203.2	89.3	235.7
2005	39.6	199.4	97.1	273.2
2006	37.8	206.1	110.4	288.7
2007	42.1	203.5	105.9	305.9
2008	88.1	297.2	149.7	497.4
2009 (Non-ARRA)*	119.9	339.3	148.1	550.2
2009 (ARRA)*	22.7	57.9	29.1	88.1
2010 (Non-ARRA)*	125.5	340.8	175.3	569.6
2010 (ARRA)*	39.7	73.6	19.6	74.2

*Refers to the American Recovery and Reinvestment Act of 2009 (ARRA), H.R. 1, 111th Cong. (2009), http://thomas.loc.gov/cgi-bin/query/z?c111:H.R.1:.

Source: National Institutes of Health, Stem Cell Information, "NIH Stem Cell Research Funding, FY 2002–2012," http://stemcells.nih.gov/research/funding/pages/Funding.aspx.

with only a few stem cell lines. By 2010, under the Obama administration and with funding flowing and advances being made on several fronts, funding for human embryonic study increased substantially. Yet non-human and non-embryonic research also continues and also receives considerable funding. This investment shows a commitment to the view that diversity and the comparison of results across different kinds of stem cells will help yield results more effectively than putting one's eggs into just one or a few baskets.

The science of regenerative medicine is grounded in understanding the basic biology of regeneration. Not all the funding for stem cell science directly supports the clinical and applied work. Stowers

Institute for Medical Research researcher and Howard Hughes Medical Institute investigator Alejandro Sánchez Alvarado, a young, highly creative scientist who is passionate about his research, realizes that we need to understand the underlying regenerative processes if we are going to be able to achieve some of the hoped-for applications. To this end, he studies the planarian, a flatworm. Planarians are great at regenerating, which is why Thomas Hunt Morgan included them in the menagerie of organisms he studied for their regenerative abilities. Sánchez Alvarado likes to point back to Morgan's regeneration studies, especially to Morgan's work on planarians in which he cut planarians apart and put them back together in different ways, even sticking two pieces together under particular conditions and generating two or even more heads.

Sánchez Alvarado refers back to this history in noting that researchers have studied regeneration for centuries; after all this effort, he asks, why do we still not understand the mechanisms of regeneration better than we do? We are not going to learn enough by looking only at humans and mice; as Morgan realized, we learn best by comparing different organisms. In some animals, we see regeneration after traumatic injury, and those are the most obvious cases. But all of us have cells dying and being discarded constantly. Despite the fact that the trillions of cells in the body are frequently replaced, the system works to regulate the process and produce a kind of tissue homeostasis. So even though the body constantly replaces those parts, the resulting organism retains its identity and individuality because of the organization of the whole system. We still think that we are ourselves, and our family and friends are still themselves, even as our cells and parts die and others replace them. All of these developmental processes contribute to regeneration and help show how successful transplantations and the right mechanisms can lead to successful regenerative medicine.

For Morgan, the key question concerned how the structure of the pieces work together, how cells might allow a planarian to change from one head to two or more heads. For Sánchez Alvarado, the question concerns what mechanisms direct the process, and in particular what molecular mechanisms. What genes are activated under which circumstances? And what cascade of gene activations leads to normal development and to the differences in these abnormal examples? Drawing on knockout techniques, using interference RNA (RNA that silences genes), and other molecular tools, Sánchez Alvarado has begun to show the causes of the developmental pathways. He also points to accidental forces in science, as when a colleague, Andy Fire (formerly at the Carnegie Institution of Washington's Department of Embryology), urged him to use RNAi techniques, which turned out to be an extremely productive line of research.[42]

This example does not just tell us about these particular worms or this particular researcher, but about the directions that such research can take. Planarians share most of their genes with other animals, including us. This, in turn, reminds us of Davidson's study of gene regulatory networks and the pathways involved in differentiation. To really access the rich diversity of stem cell applications that investors, the public, and researchers hope for, we will need a careful study of genes and evolution, the molecular mechanisms involved, and the phenotypic patterns of cellular and tissue structural changes.

So we see that regenerative medicine starts with the foundational basic study of stem cells and how they behave under different conditions, but it does not stop there. In order to understand how to construct effective regenerating systems and to develop effective use-inspired research, we need to understand systems and their interactions. We can understand those systems best when we get at the mechanisms underlying them, which in turn involves learning about the particular genes involved. Also, as researchers have shown repeatedly, we can learn a great deal about complex systems by analyzing diverse model systems rather than just one or two. Only then can we begin the process of synthesizing effectively—with greater understanding about which times it will be biologically possible to do this at all.

In the absence of clear guidance or regulation, amid deep and powerful convictions about what is right and true, and with new scientific and technological innovations appearing in areas like stem cell research and regenerative medicine, the diverging views about embryos bring clashes. We can see how such conflicts arise, and we can see why doing nothing federally is the easiest path. Yet states have not been as reluctant to weigh in on embryo-related issues. The present situation, with increasing political pressure to restrict research as well as reproductive rights in some states even as other states vote to use public funding to develop embryo-based medical treatments, is not stable. Nor does this uneasy current situation fit well with the long-standing contract between science and society that calls for public investment to generate knowledge that is applied for the public good. In Chapter 8, we look forward at current and emerging areas of biological discovery that are likely to raise further questions. As basic research continues to turn toward clearly useful applications, however, perhaps some of the current tensions will give way.

8

Constraints and Opportunities for Construction

What we know about developing embryos has increased tremendously, with increasing speed, as we have seen in the previous chapters. Stem cell research and regenerative biology generally are already leading to use-inspired research and medical applications. Yet the logical extension of Jacques Loeb's vision of not just understanding but actually controlling life is only beginning to come into reality. So far, we have seen a great deal of observing, experimenting, using microscopes to see more detail, extending research on cells into glass dishes with culture media, and imagining ways to use stem cells to replace lost function.

Loeb would probably urge us to think further. Why not synthesize life, or at least some parts of a living organism? Synthetic biology can start with already living building blocks and reengineer them by constructing new parts, or it might involve creating life from nonlife. This chapter looks at examples of synthetic biology: tissue synthesis using cells to reconstruct missing parts, synthesizing cells from biological parts, and the idea of synthesizing embryos. Along the way, we have to consider issues related to patenting and who owns the products that are created: is it possible to patent or to own life?

The Concept and Context of Synthetic Biology

Synthetic biology is relatively new as a recognized field, even though the idea of controlling life and even engineering it in various ways has been around for a long time. The words "synthetic biology" now bring up millions of hits in Internet search engines. The histories by current researchers typically point to the 1970s and the molecular successes with recombinant DNA techniques as the starting point for thinking in these terms. At the 1971 meeting of the American Association for the Advancement of Science, the session "Living Systems: Synthesis, Assembly, Origins" was an example of an early discussion of the ideas.[1] During that symposium, the speakers pointed to new possibilities and also raised questions about whether there should be limits to the work done, which signaled the beginning of new thinking about synthesizing life. This rebirth of interest at first largely focused on molecular assembly—that is, on taking molecular parts and putting them together in engineered ways.

By the first decade of the twenty-first century, the term had expanded to include several additional types of synthesis. A number of international conferences called for further investment and research into synthetic biology; at the same time, calls arose for reflection on the ethical and social justice questions involved. By this time, recombinant DNA and molecular assembly still played visible roles, but the field now included techniques for synthesizing new molecules and for synthesizing new strands of DNA. Researchers began to discuss the possibility of actually making—not just imagining making—a synthetic cell. Others began to envision synthesizing replacement parts such as skin tissue, organs, and other living structures.

Synthetic biology has come to mean work at the intersection of the study of biology, with engineering, computation, and modeling to synthesize something that actually works. The goal is for the resulting engineered product to function successfully as "life," and for the work to count as "biology" and not merely as engineering using only inorganic material. That is, it is one thing to make an artificial part from nuts and bolts and plastic or titanium, but quite another to make it from cells and tissues and in ways that integrate with living material. There are limits to the first kind of engineering, and the dream for synthetic biology is to achieve the latter. This is not just because scientists love to know things and want to be able to control life for reasons of curiosity. Rather, a major motivator is the use-inspired desire to help solve medical problems and to make life better for those experiencing the functional limitations of missing parts.

Sometimes the practical goals may not be clear, and it may seem that researchers go too far. When the public is confronted with a new discovery, without context and without explanation of why the research was done, questions naturally arise. This happened on May 20, 2010, when Craig Venter announced that his laboratory had created a cell.[2] In fact, his cell had begun with already existing life components, so the laboratory did not fully create new life from nonlife. Nonetheless, his announcement triggered vigorous debate about whether there should be limits on engineering life. In response, in the United States, President Barack Obama asked his Presidential Commission for the Study of Bioethical Issues to look at what Venter had done and at the ethical questions raised by the work.

The commission issued its report, "New Directions: The Ethics of Synthetic Biology and Emerging Technologies," in December 2010.[3] This excellent, clear report emphasizes the importance of working within ethical boundaries to maximize public good while minimizing risks. The report points to five traditional ethical principles as driving decisions about synthetic biology and its related technologies. These include the goal of providing public beneficence, exercising responsible stewardship over resources, balancing intellectual freedom and responsibility, relying on democratic deliberation about decisions, and promoting justice and fairness.

Because so many different things have been thrown together under the banner of synthetic biology, it is worth reviewing what was taken to be at issue for that report. Given that the commission clearly spent a great deal of effort working on the precise wording and providing the launching point for discussions that have followed, it is also worth reviewing the definitions in their own words. In particular, the commission defined its subject matter in the following way:

> Synthetic biology is the name given to an emerging field of research that combines elements of biology, engineering, genetics, chemistry, and computer science. The diverse but related endeavors that fall under its umbrella rely on chemically synthesized DNA, along with standardized and automatable processes, to create new biochemical systems or organisms with novel or enhanced characteristics. Whereas standard biology treats the structure and chemistry of living things as natural phenomena to be understood and explained, *synthetic* biology treats biochemical processes, molecules, and structures as raw materials and tools to be used in novel and potentially useful ways, often quite independent of their natural roles. It joins the knowledge and techniques of biology with the practical principles and techniques of engineering. 'Bottom-up' synthetic biologists, those in the very earliest stages of research, seek to create novel biochemical systems and organisms from scratch, using nothing but chemical reagents. 'Top-down' synthetic biologists, who have been working for several decades, treat existing organisms, genes, enzymes, and other biological materials as parts or tools to be reconfigured for purposes chosen by the investigator.[4]

The report thereby reinforces the point that different things fall under the synthetic biology banner, which will likely come to include still others that we have not yet imagined. Initially, work on recombinant DNA offered ways to take the DNA from one organism and even one species and combine genetic material with that of other organisms or species. The report's discussion of the science starts there. It then proceeds to show how Venter's group had used similar and additional techniques for their production of a constructed cell.

What is next? the commission asked, and they acknowledged that in such rapidly developing fields it is impossible to know. Similarly, it is difficult to predict how synthetic systems will behave. Will they act exactly like the natural parts they are intended to mimic or replace? Or will they begin to act differently in different environments? This question is especially important if synthetic parts constructed in the laboratory are placed into people. We know that the culture medium matters, so we should realize that a change of environment will matter. How to take this into account and test carefully for safety raises many practical as well as ethical questions, as the report notes.

We can tell that synthetic biology has become mainstream and is not just a wild-eyed fancy by extremely creative and provocative scientists like Craig Venter by looking at the International Genetically Engineered Machine Foundation (iGEM), a group that focuses on synthetic biology. As the group's website (http://igem.org) explains, their idea grew out of a 2003 course at the Massachusetts Institute of Technology (MIT) and has become a widespread movement "dedicated to education and competition, advancement of synthetic biology, and the development of open community and collaboration." High school and community college competitions join with more senior scholars in developing new projects. The results demonstrate just how much innovation emerges when creative energy is focused on the collective goal of creating and sharing ideas and technologies to produce engineered biological components.

The process has brought a number of inventions into production. And, again, it is important to recognize that this is not wild-eyed science in the absence of social concern; the movement is entirely dedicated to bringing the power of biological and computational tools to promoting the public good by solving medical problems. The iGEM group wants to work openly and share its ideas and results. Some lines of research in synthetic biology just beginning now are building on the knowledge already gained, and they are likely to play out in new ways over the next decade. Some ideas are moving into the production of applications; others are still under development. One challenge concerns who will own the results, and we need to look at patent policies before proceeding.

Patent Policies

Chapter 7 reflected the upbeat feeling of the stem cell research community that they are making progress, and this chapter shows the enthusiasm among synthetic biologists. They are clearly putting together pieces of the puzzle that will help them gain the knowledge to eventually engineer cells to produce therapies. In recent decades in the United States and much of Europe, researchers have had the right to file patents to protect their inventions and their intellectual property. The underlying assumption, which has governed patent law for centuries, is that an inventor deserves to have protection for the invention so that others cannot steal the idea and develop it themselves. Genetic engineering has challenged that time-honored assumption in new ways with the question of whether identification or even manipulation of genetic information is actually an invention.

The core question is whether it is possible to patent life forms, and if so which ones, under what conditions, and with what limitations. Billions of dollars are at stake when a pharmaceutical company develops a genetic test, for example, and wants to control it completely. Similarly, patenting stem cell lines or technologies could involve large sums of money as well as raising important questions about who controls the research. There has been a somewhat unsettled consensus that some kinds of biological tests and information can, in fact, be patented while others cannot. Yet each country has its own restrictions and permissions, and the situation to date has involved an uneasy set of agreements.

October 18, 2011, a ruling by the European Court of Justice appeared to deal a serious blow to such research. Oliver Brüstle, a neuropathologist from the University of Bonn in Germany, was defending his right to file patents for his human embryonic stem cell research. The Greenpeace organization had attacked the claim that a researcher could patent life in this way. To the surprise of Brüstle and most others in the research community, the European court upheld the lower court's ruling in favor of the Greenpeace restrictions. It went much further and also ruled that research on human embryonic stem cell lines was immoral. This was not only a judgment of morality but also about the right to patent human cell lines.

The initial reaction by the press emphasized the loss for stem cell researchers and raised questions about just which research should be allowed and which should be prohibited.[5] Debate has ensued in the various countries' legislative bodies as well as among the European Parliament. Past debates led to acceptance of human embryonic stem cell research and even to allowing funding for the work. But the court's ruling reopened the debate. Clearly, the legal and regulatory context in which stem cell research takes place is likely to continue evolving, and some supporters feel that making stem cell lines unpatentable may actually open researchers to a greater exchange of ideas and materials. As decisions about patenting genetic sequences continue to undergo rethinking and revision, the resulting rulings are likely to impact thinking about other patenting of information about biological processes.

In the United States, we have yet to realize the full impact of the June 13, 2013, ruling by the Supreme Court in *Association for Molecular Pathology, et al. v. Myriad Genetics, Inc., et al.*[6] In this case, the Association challenged Myriad's right to patent human genes or DNA sequences that occur naturally. On the face of it, this ruling taken in the context of other rulings, could serve as a significant obstacle for companies wishing to invest in development of stem cell therapies. Yet the courts have allowed the patenting of processes, and companies have found ways to protect their investments. The legal status of stem cells, stem cell lines, stem cell culturing techniques, and the products resulting from using naturally occurring stem cells have not yet been fully determined. We are likely to see continuing discussions of these issues. And the legal and regulatory contexts will undoubtedly affect what research actually gets done and by whom. Nonetheless, synthetic biology proceeds.

Tissue Synthesis

One type of engineering approach involves the kinds of tissue engineering we have already discussed. Earlier, we focused on the possibilities and on science that was already reasonably established. This chapter looks at the work under way right now and points to the future. Medical problems motivate research on tissue engineering, specifically missing or failed tissues and organs in the body. In some cases, the failed or missing body part performs relatively mechanical functions, and it is these parts that offer early promise for successful replacement. Here, we look at a selection of particularly intriguing examples.

Even the name of the trachea (or windpipe) suggests an engineering function. In 2008, researchers in Spain first succeeded in using stem cells to construct a tube to function as a trachea for a patient with tuberculosis. As a result, the thirty-year-old mother of two delightedly declared that she had regained her life.[7] In this case, the researchers employed their knowledge of how cells work, derived from the historical study of cells and development, and added that to mechanical engineering knowledge of how the tracheal structure performs its functions. They built a material scaffold of the right size and shape, then added living cells to craft a functional structure that worked both in the laboratory and when placed inside the body.[8] Although the process may sound relatively simple, each of those steps involved a tremendous amount of research, testing, and skill combined to produce the successful synthesis of a new trachea.

In January 2012, researchers in Sweden announced that they had gone still farther by adding a patient's own stem cells to a plastic trachea.[9] A male patient in Maryland who had developed cancer of the trachea was considered inoperable because simply removing the diseased tissue would not solve his problem—only a replacement would work. Transplanting tracheas from a donor is essentially impossible because few are available, they must be the precise size and shape, and the body immediately tries to reject the foreign object. They needed to replace the structure with one that the body would accept. Their solution was to design a scaffold from nanoengineered plastic and to seed the plastic with the man's own stem cells taken from his bone marrow. They placed the structure with its stem cells into a bioreactor, which creates an environment in which the stem cells can multiply and infuse the plastic. The cells undergo repeated divisions in their new environment and produce cartilage. In effect, they respond to the biological conditions and produce the type of tissue needed, presumably because of internal cell signaling. This is a marvelous example of controlling life, and surely Jacques Loeb would have been thrilled.

Subsequent cases have shown the capacity to grow a new trachea in young children by seeding the right area with the right kinds of cells in just the right environmental conditions, and experiments are under way for other organs. These examples of using existing cells and their biological capabilities to combine with engineered inorganic materials suggest that there may be tremendous potential for additional extensions into similar kinds of work. Such new technologies are making these approaches less about the basic mechanics and more about using bioengineering to bring together the intersections of organic tissue and mechanical structures and processes. This allows not just replacing the workings of a part but doing so with something much more like the normal structure and function.

A few additional examples give a preview of what more is likely possible. In June 2012, researchers—again in Sweden—reported a successful transplantation of a vein grown from a patient's own stem cells. The ten-year-old girl in Sweden suffered from blockage of a vein to her liver, and the traditional medical options included a liver transplant along with reengineering to allow blood flow, but such transplants carry risks and require finding a donor. Another option would have been to take one of the girl's own veins from another part of her body and transplant it to bypass the blocked vein, but that also had considerable risks and raised special problems for one so young. Instead, they decided to try a new approach. They took a vein from a dead donor and cleaned it of all the cells that might cause an immune reaction. This provided an appropriate scaffold. They then cultured stem cells from the girl herself, taken from her bone marrow, and let those stem cells grow and infuse the vein scaffold—just as the researchers had done with the trachea case.

This approach has such potential that researchers have had some success with other structures as well. Again, the combination of material scaffolding to provide the structure along with stem cells to provide the function—the stem cells taken from the patient him or herself (and hence what is called an autologous transplantation)—provides a system that works for these cases of replacing damaged parts.

Even more exciting is the next step of using stem cells for therapies that cause the cells to differentiate when they do not have a mechanical scaffold. An announcement in June 2012 at the International Society for Stem cell Research created a stir. Researchers at the RIKEN Center for Developmental Biology in Kobe, Japan, led by Yoshiki Sasai, had taken human embryonic stem cells and cultured them, as many researchers have done in hopes of getting them to produce useful tissues.[10] Sasai's team knew that reproducing normal conditions that cause the stem cells to differentiate is very difficult. Most stem cell cultures multiply cells and lead to a two-dimensional sheet of cells, but Sasai's laboratory transcended this previous limitation. They had already shown with mouse stem cells that they could produce three-dimensional structures that were organized as cerebral cortex and as precursor to the pituitary gland. When they attempted to get human cells to form into an optic cup, they succeeded.

As Sasai explained, research on the optic cup has a long and rich tradition, going back to the transplantation work of Hans Spemann. Their experiment produced a cluster of epithelial cells that formed into a ball and bulged in a way that then folded back on itself. This formed a sort of pouch, which led to the structure of the optic cup. They succeeded with more than just the structure, though; the cup contained the different kinds of cells that make up the normal inner and outer layers. What they had shown is that complex differentiation is possible, and also that the experimental culture conditions could reproduce a structure that looks normal.

This was just another early step in achieving clinical applications, but the research is very promising. As of late 2013, others have not succeeded in replicating their results, but many are trying. Some clinicians have begun to experiment with layers of cultured retinal cells as well. Because the human retina is relatively tolerant and does not invoke a violent reaction to foreign cells, such retinal therapies may be easier to achieve than with tissues that have a stronger immune response.

Also in June 2012, Song Li at the University of California, Berkeley, announced results of his studies of the vascular system.[11] We know that the blood vessels carry stem cells, and also that they are the source of additional stem cells. Yet the role of those cells has remained unclear, and the enthusiasm for stem cell therapies has tended to emphasize their positive roles rather than possible problems. Li's team showed that the vascular stem cells may be responsible for some kinds of vascular disease. Their research performed in mice has implications for our understanding of human diseases as well. Some commenters see the potential to turn this new understanding of the possible negative side of stem cells into knowledge about how to control the cells to prevent or treat disease. In fact, an Australian team from Sydney and New South Wales led by Richard Harvey has discovered a new type of stem cell in mouse hearts that holds promise for repairing loss of heart muscle function.[12] The negatives and positives do come along together.

Treatments using these new engineering technologies depend on successful tissue synthesis, which has also received considerable attention. In 2012, for example, the March of Dimes had presented their annual Prize in Developmental Biology to two researchers for their study of skin regeneration. Howard Green at Harvard Medical School, and his former postdoc Elaine Fuchs who had become a Howard Hughes Medical Investigator at Rockefeller University had discovered much about the cellular actions underlying the causes of skin diseases.[13] Their discoveries have stimulated work on wound repair that picks up on the themes of regenerative biology initiated by Thomas Hunt Morgan and carried on by Ross Harrison, among so many others.

In July 2013, researchers in Japan announced that they had produced the first steps toward synthesizing a liver.[14] When human stem cells cultured to grow into liver cells were placed in laboratory mice, the cells grew in this new environment, differentiating as liver cells up to what is called the liver bud stage. Although the cells did not go on to produce human livers inside the mice, the technique and results show how far these researchers have progressed in understanding the developmental processes and in being able to control differentiation.

Another area where there is great hope but so far less progress is with the pancreas. Diabetes, a widespread disease, involves the inability of the pancreas to produce insulin. Treatments to date have largely depended on replacing the organ's function with substitute or synthetic insulin, and researchers have long sought more effective replacements. The ideal is to create a way for the body to produce its own insulin, which would require the body to produce sufficient cells called beta cells. Patients with type I diabetes lose their beta cells until none are left; type II patients typically retain some function. In both cases, increasing the number of such cells would help treat the disease, and in the case of type II diabetes, increasing the capacity of the cells to produce insulin would help. To date, the efforts to cause the body to produce more beta cells or to transplant the cells effectively have been unsuccessful. In this case, the complexity of the body as an adaptive system has put many obstacles in the way, but the stakes are high. In mice, researchers have made some progress in understanding the complex system and in engineering solutions. Reports from late 2012 that indicated that researchers have identified and isolated pancreatic stem cells and caused them to begin some function offer increased promise.[15]

Another quite different example of deriving and using information is the "lung on a chip" developed by Harvard researchers in 2010, which has been improved since then.[16] They used computational methods to provide a biomimetic stand-in for the human lung. The device they engineered is not a mechanical lung, and it does not look like a lung; rather, it is a computational device designed to mimic human lung function, to behave in ways parallel to the way that lungs actually work, making it useful for medical testing of drugs and procedures. The constructed "lungs" are not intended to replace the lungs in the body, but they could lead to knowledge about how to do so. In the meantime, they provide a way to test how diseases impact organs. Donald Ingber, leader of the research group, reports that they are working toward similar biomimetic devices for the kidney, heart, lung, gut, and others tissues.

For still other lines of research focused on regenerating biological parts, leading researchers in regenerative biology such as Sánchez Alvarado estimate that within five years they will have the gene activation cascades worked out. Others believe that will happen even sooner: by some estimates, within five years some stem cell lines will be sufficiently well understood that we will be able to map out clinical applications for regenerating function in very severe cases of injury or disease. Others believe that we are close to understanding a wider range of gene regulatory networks in organisms beyond Davidson's purple sea urchins, which should make it possible to begin managing the systems more effectively.

Nobody believes we are close to being able to cause cells to differentiate predictably and safely in ways that will allow them to be placed into the body and reliably continue to do exactly what we want. Nor can researchers create organs such as tracheas or eyes or hearts from scratch. Yet they can take away some cells or add others, and they can recombine, knock out, or add genes. If optimists like Sánchez Alvarado are right, soon we should be able to change the gene activation sequence in a variety of ways. At that point, we can take the stem cells out and culture cell lines for research and quite probably for at least some therapies. All of this is possible with existing cells, which we can manipulate and control to function as we want.

Making Cells from Scratch?

Our next question, then, is whether we can go even further—by creating cells. Enter Craig Venter, well known for being brash and bold and taking on what others consider impossible tasks. With the Human Genome Project, Venter felt that the federal government's sequencing project, funded through the National Institutes of Health (NIH), was taking too long and using a very inefficient approach. So he started his own company to develop alternative methods, and challenged the government to a race to finish the genome—forcing the NIH to work faster to keep up.[17] The comparison of different sequences helped the entire process become both faster and also in the end more accurate and useful.

As mentioned earlier, Venter's May 2010 announcement that his group had developed what they called a synthetic cell led to President Obama calling for a study of the ethical implications of such synthetic work.[18] Venter had spent about $40 million and developed a simple one-cell organism with the capacity to reproduce itself. In fact, they started with a cell from which they had removed all the genetic material—as we have seen before with nuclear transplantation experiments. But in this case, they did not transfer in another nucleus from another living cell but took the largely empty cell and gave it new DNA. Notably, and the reason this work was considered a breakthrough deserving special attention, the DNA did not come from another organism but rather from synthetic sequences created in the laboratory. Of course, they did not choose any old random sequences; instead, they used synthetically generated sequences that followed patterns of natural bacterial species.

Did Venter's team create life? No, not really. Did they create an artificial cell? No, not fully. But they did create something monumentally important nonetheless: a DNA sequence that they could put into a cell that then replicated and did cell-like things. Their step toward synthetic cells shows that the genetics part is possible. Now it remains for researchers in what is called systems biology, or the study of how complex systems work, to understand the complexity involved in creating a completely artificial cell. Will it happen someday? Likely, yes. Will it happen soon? Likely, no. The reactions of researchers and bioethicists to the announcement varied considerably.

More recently still, a group of researchers centered at Stanford and including others from the Craig Venter Institute in Rockville, Maryland, announced that they had constructed a whole-cell computational model to predict the phenotype from a genotype.[19] The team, led by Jonathan Karr and Jayodita Sanghvi, relied on bioengineering and biophysical expertise to model how a genotype can give rise to the complexity of the phenotype. In particular, they modeled the human urogenital bacterial parasite *Mycoplasma genitalium*. This bacterium, which has only 525 genes, was much easier to work with than a common human bacterium such as *Escherichia coli,* which has an estimated 4,000 genes. The research team announced that they had used compiled databases to draw from over 900 publications and had used over 1,900 parameters and many other factors that include details about the genome, proteome, transcriptome, and metabolome. What had once been a science-fiction concept of creating life has begun to move from theory to practice, a computational problem tractable using available technologies.

The emphasis on synthesizing a whole cell, rather than parts that must be assembled, suggests a future effort to synthesize whole organisms. If we can create a cell, why not synthesize multiple cellular structures that also involve cell-cell interactions? Multicellularity involves an additional complication because the way cells interact changes over time. At the embryonic stage, cells remain within a protected environment. That is, the human embryo is safely protected inside the uterus, where it develops and grows. Cells divide and differentiate, organs emerge and begin to function, the embryo becomes a fetus, and the fetus is eventually born and enters a more complex world. This entire process is protected by the placenta so that the individual organism is largely separated from the larger environment. Nutrients and waste move from the mother to the developing embryo/fetus, and cells move across the placental barrier. But, on the whole, the developing embryo and fetus are insulated.

After birth, a human becomes a complex microbiome, a community of microorganisms all living together. And cell-cell interactions are influenced not just by the neighboring cells, but by the microbiomic environment. Mimicking and replacing adult function will involve understanding the role of all the parts, including the microbes, and that research has only just begun.

The microbiome begins when the fetus is born, either by moving through the birth canal or extracted from the uterus via the surgical procedure called a cesarean delivery. Especially during a normal birth, the fetus absorbs bacteria and other microorganisms, and at this point the individual human microbiome begins in earnest. We tend to think of ourselves as being individual, autonomous organisms that began with an egg and sperm coming together. We are that, but we are so much more. Recent estimates suggest that no more than 10 percent of our cells are actually human—the rest are microorganisms that are part of the vast community we carry around. Researchers are just starting seriously to study the ways that this microbial world shapes our development. Or, in the view of developmental biologist Margaret McFall-Ngai, researchers have stopped ignoring the elephant in the room in their focus on cells, fertilization, differentiation, and the other traditional developmental topics, and are beginning to take into account the microbial community.[20]

This has led researchers to seek to identify and sequence the DNA of all those other parts of us. The Human Microbiome Project is an ambitious effort with the goal of identifying the estimated 100 trillion "good" bacteria that live within each of us. "Probiotic" diets draw on the assumption that we know what is good and what is not; and some patients find themselves in what sounds like the odd situation of taking probiotic and antibiotic medicines at the same time. All this effort is aimed at getting the system in balance, although we still have relatively little understanding of what that means.

As Leroy Hood, the founder of the Seattle-based Institute for Systems Biology, noted in an editorial calling for support for the microbiome project, we know astonishingly little about the components of the system or their interactions.[21] Yet we have a lot of data, and as Hood said, "The ultimate goal is to convert data into knowledge." With knowledge, "it may be possible to reengineer microbial genomes to make them execute the chemistries that we desire." We can do that by hunt-and-peck approaches or by using effective computational tools. He concluded, "For success, academic and industrial research institutions will need to create the focused, integrative, milestone-driven, and cross-disciplinary environments in which transformational systems-driven innovation of metabolomes can thrive." Understanding organisms starts with embryos, but actually constructing replacement structures and functions moves to a vast community of individuals working together, with each influencing the development and life of others.

Synthesizing Embryos and Creating Life?

We remain a very long way from constructing embryos synthetically, and it is not clear why we would want to create embryos completely from scratch. Yet, as previous chapters have shown, we know a great deal about how embryos develop, even though much remains to be learned. Hans Driesch still inspires us with his emphasis on the marvelous regulatory capacity of cells and embryos to respond to changes while still retaining the integrity of the complex interactive whole. To model and then create a living whole embryo seems like a daunting task.

Yet we have seen that it is already possible to manipulate embryos. In humans, fertility clinics remove one or two cells from the eight-cell stage for testing, a standard practice when genetic testing is needed. The remaining six or seven cells respond and develop perfectly normally, as Driesch might have predicted they would. We therefore accept the value of taking the embryo apart in this way for a useful application.

We also have the technology to modify embryos genetically. In mice, as Mintz and others showed, it is possible to put together cells from different embryos or even whole embryos and they develop into normal mice. We can imagine using such chimeric combinations to solve genetic problems. For example, if a couple who wants to have their own genetic children can only produce embryos that lack certain genetic factors, in theory a clinician could add cells that carry those factors. The resulting child would be a genetic chimera, or mosaic, in which some cells in the body carry that gene or genes and others do not. The cells that do carry the gene could perform their necessary functions for the child as a whole.

This kind of embryo reconstruction can happen in theory, and probably also in practice judging from work in mice. But now we reach an area that reasonably evokes ethical debate. Why would we want to do this kind of engineering? We can ask the same questions for other aspects of life engineering, and the answers must come from social decisions rather than strictly biological answers. Yes, such manipulations may be technically possible, but is there any legitimate reason to do them? What use-inspired goal does such research serve? Are there limits to scientific and technical knowledge and applications? Yes, of course, there are—but what are they?

Adding Social Questions to the Biological

The story of the mythical Prometheus and his suffering as his liver regenerates every night could be a guide for thinking about our conflict about what is considered natural, what is considered a legitimate intervention in treating medical problems, and what is considered going too far or transgressing some perceived boundary. When is it that we go too far? Are there times when we do not really want scientific discoveries and applications? Or are we constrained only by our own imaginations and possibilities?

Dr. Frankenstein is generally thought to have gone too far—or that is what we are supposed to feel. Other science fiction stories also have presented morality tales of carrying human exploration and intervention into nature too far, with dire consequences. See, the message goes, if you transgress natural boundaries and try to control nature, you will fail. Even when your intentions are good, you are likely to do more harm than good because of your hubris—thinking that you know more than you do, that you can control what you cannot.

Yet the other side, also captured in science fiction and futuristic writings, is the progressivist conviction that science and technology can, in fact, consistently make things better. This was Isaac Newton's and Francis Bacon's view in the seventeenth century, and it was the attitude during the Scientific Revolution generally. God had made the natural world, these leading thinkers believed, and God had also made us and given us the powers to understand and interpret the world. Thus, he clearly intended for us to use those powers and to turn them to doing good. Use-inspired science is the way to progress and to overcoming human misery. The social contract for science by the mid-twentieth century insisted that governments should invest public funding for scientific research in order to harvest the applied results. But how? And how far?

Again, what is the appropriate social contract for science? If we agree that use-inspired research makes sense, who decides which uses and what outcomes we seek? We have reached the limits of our ability to answer such questions based on biological understanding alone. We need to draw on our historical perspective and look at the implications of the diverging meanings of embryos and of life.

Therefore . . .

This book recounts a history of looking at embryos through microscopes and other techniques, and thereby gaining a growing understanding of their biological reality. Looking at the embryo reveals a complex interactive system, which adapts to changing environmental conditions. It is worthwhile to review the biological details here quickly, because getting the facts right is very important for grounding policy decisions. Even though much of the history presented in this book has been drawn from the study of a wide variety of animals, let us now focus on humans alone.

Each cell is highly structured: the cytoplasm surrounds a nucleus that contains inherited material carried in the DNA of the chromosomes. Cells interact, and the signals among them and with the surrounding environment help guide the regulation of gene expression. The fertilized egg cell divides into two cells, then four, then eight, and so on in more complicated ways. At first, the cells just divide physically, and the embryo does not grow any larger. Although the embryo begins to consist of a larger number of cells, these are essentially of the same kind and are not yet differentiated or undergoing gene expression. As we shall discuss, this earliest form is often called the preimplantation or even the preembryo stage.

The blastocyst stage starts the beginning steps toward the process of development and differentiation. At this point, in humans the blastocyst is a single layer of cells surrounding a partly hollow center. Inside, the inner cell mass contains pluripotent human embryonic stem cells. Those special cells have the capacity to differentiate into any of the diverse kinds of cells that make up a complex differentiated organism, depending on the conditions, but none of them can become an entire organism. That is, they are pluripotent but not totipotent. By the end of the blastocyst stage, at five days to no more than fourteen days, this (pre)embryo must be implanted into the mother to survive and continue its development.

The microscope and other experimental techniques have shown us all these details and more, and researchers have observed many embryos, which can survive in a glass dish up to the blastocyst stage. In fact, many embryos do live in glass dishes in fertility clinics, produced for otherwise infertile couples who are eager to have children with their own eggs and sperm. Thus, since the beginning of in vitro fertilization in 1978, we have learned a great deal about these early developmental stages by direct observation. However, studying the developmental stages after implantation requires other techniques—indirect observation, the study of other animals, and inferences from the accumulating evidence. The latter research involves looking at living embryos in mice especially, which develop in ways similar to humans. Such experimentation involves employing control conditions and discovering which variables cause which results, and with these methods researchers have learned a great deal.

The accumulating evidence has made clear that in the beginning the embryo is not yet formed. In other words, it does not yet have structure other than as clusters of cells, and it does not yet have any of the functions—such as eating, digesting, breathing, or sensing—or other activities that we typically think of as defining a living human organism. Yet the embryo is not wholly uninformed, and it is the inherited information that helps to guide each cell to differentiate from pluripotent embryonic stem cells that could become any kind of cell into becoming particular, specialized kinds of cells. This differentiation, which happens gradually over time, developmental stage by developmental stage, results from the inherited material in the DNA in the chromosomes interacting with both the environment of the cell itself and the environment external to the cell to direct what is called gene expression.

The process leads, again gradually and stage by stage, to what is called a fetus. At this point, at about eight weeks in humans, the organism has a rough structure laid out, and all the organ systems are present in rudimentary form. Gradually, a more detailed structure continues to develop, and function will begin. At the time of birth, the organism will be a functional whole. Along the way, a lot can go wrong—and it often does—yet the embryo at its earliest stages has the ability to respond to challenges and to regulate and adjust.

Biology is exciting and informative, giving us insight into the beauty of life. What we see with the microscope is the amazing developmental process by which form emerges out of unformed material. To watch an embryo develop is to see the life of a new individual emerge, gradually, stage by stage, from very simple material that started with the coming together of two cells. Even though we cannot watch every step of human development, because the embryo is inside the uterus, we can see these steps in other organisms and can use inference to understand human stages as well.

Those who have not had the privilege of watching embryos develop in the laboratory often have learned about the process through television specials, videos, the Internet, and print images. Yet some people choose not to look into the process at all, and as a result they miss the chance to see the wonder of biological development and opt for a limited meaning of life. Those who consciously choose not to see what a material embryo is—instead relying on metaphysical assumptions and their imaginations—miss the astonishing ways that the embryo responds to environmental conditions. In imagining a preformationist process in which fertilization somehow creates a miniature organism that follows a simple process of growth, they are missing the intricacy of development and differentiation. For those who seek mystery and beauty, both are abundant in the ways that developmental processes unfold.

Thus, those who have looked through a microscope (either in person or indirectly by means of images or reports from others) and those who have not tend to have diverging understandings and to assign divergent meanings to life. The result may range from general disagreement to outright conflict, especially when dealing with decisions about ethics and policy.

Policy Implications for the Embryo

The study of biology cannot tell us what we should do in the world, and it cannot tell us what is ethical or good or right. But biological research does show us how the world works and tell us what can be seen, what are the facts, and what counts as evidence. A rational and enlightened society will demand that its decisions about policy should be at the very least informed by and consistent with scientific facts. Making social decisions that contradict biology, in this case, simply does not make sense—the world is not flat, humans do cause global climate change, the evolution of species with humans included does occur, and an embryo is not a fully formed human organism. We start with facts, and then we consider their consequences for a couple of hot button political issues. In particular, we know that fetuses before the third trimester do not feel pain. Therefore, it should now be clear that it makes little sense to designate the unformed cluster of cells in the dish as a "person" in any biological or robust legal sense.

One thing we learn from looking at the embryo is that it goes through stages of development, each of which is biologically very different, and in which form and function emerge only gradually. Even though it is the same organism that begins with fertilization and develops eventually into an adult individual, the stages along the way are very different. The developing individual has continuity, yes, but it is not at all the same kind of thing at every stage from fertilization to adulthood. Instead of lumping all the stages together, then, it makes sense to look at them separately and to consider whether they might have different meanings socially as well as biologically. Our policies should be informed by this understanding.

The first stage is that from fertilization up to the blastocyst stage, or the time of implantation into the uterus. Following his advisors at the National Institutes of Health, United States President George W. Bush called this the "pre-embryo" in August 2001 when he gave his address about embryo research.[1] Setting up a distinction between the preimplantation pre-embryo and the postimplantation embryo makes sense because tremendous developmental changes occur at those times. We can directly see the preimplantation embryo in a fertility clinic or research laboratory dish. With implantation, the embryo disappears from direct view: now it is inside the gestating mother, and at that point, it becomes dependent on the woman. Now, the embryo begins to grow rapidly, to continue dividing into more cells, to undergo cell differentiation under the guidance of gene expression and response to the environment, and to begin the process of what is called *morphogenesis,* the genesis of form.

The first or "pre" stage, when the embryo is not yet undergoing any significant gene expression or growth, is a period when the material cells are really just that. Again, this undifferentiating stage lasts from fertilization up to formation of the blastocyst. These pre-embryos are really only clumps of material cells at this point. Recognizing this fact is important for political as well as biological reasons. One example of a clash of diverging meanings assigned to the embryo concerns ideas of "personhood." Biologically, it is very difficult to argue that this cluster of undifferentiated cells in a dish is a person in any other than the most simplistic sense, but that is what advocates of the personhood movement argue. Let us look at their reasoning and at how biology can help us understand these claims.

The Personhood Initiative

As is rather obvious from listening to and reading public media, the United States is home to a very active movement that opposes abortion and packages their opposition in whatever form seems to be helping their cause. By the time of the 2012 presidential election and in continuing activity focused on selected states, one branch of this movement centered around what its advocates referred to as the "personhood initiative." Personhood.net, Personhoodusa.com, and other Internet presences illustrate what is at issue. In essence, all their efforts are directed at defining embryos at all stages as legal persons and ensuring that they have the rights and protections of citizens. At the national level, this has played out as discussed in Chapter 1, when in 2012 Paul Ryan and others proposed H.R. 212 in the U.S. House of Representatives, the legislation that they called the "Sanctity of Life Act," which sought "to provide that human life shall be deemed to begin with fertilization" (or its functional equivalent through cloning or other manipulations). According to this legislation, "every human being shall have all the legal and constitutional attributes and privileges of personhood."[2] At the state level, Wisconsin, Colorado, Mississippi, Georgia, Texas, and other legislatures also have explicitly focused on the concept of personhood.

This is a telling example of how diverging meanings derived from biology and metaphysics come into conflict in the social and policy arenas. People who start with metaphysical assumptions originating from millennia ago can easily retain the ancient, hypothetical imagination of an embryo as the beginning of a very tiny person beginning at the earliest stage of development. In contrast, those who have actually seen an embryo have viewed a cluster of unformed cells in a glass dish, in which form emerged only very gradually, over time. Yet if we are going to invoke ancient ideas, then why not the Jewish or pre-1869 Catholic views, in which up to forty days the early embryo is "like water" and not yet ensouled or quickened into life? Better yet, let us remain grounded in the biological facts about what we can see, then sort out the implications.

For those tempted to regard this divergence as between science and religion, let us recall that the intellectual leaders of the seventeenth-century Scientific Revolution and the eighteenth-century Enlightenment were convinced that God had given humans the ability to use both our eyes and our microscopes. With our microscopes, we can see that embryos are clumps of material cells, and we can learn a great deal about development from studying them. We then can use the knowledge we have gained from studying developmental processes in general and stem cells in particular to develop medical innovations to help people.

The personhood movement is a social belief system and a political movement attempting to impose the group's personal views on society at large. The personhood supporters are doing their job effectively in some states and in some contexts. As the Personhood.net website puts it, "To say that these embryos are 'cellular' life but not human life is to engage in a game of semantics. Every one of us started out as embryos."[3] They want us to see an embryo as the same kind of thing as an adult human, and therefore legally the same. In much of their literature, they imply that there is a biological sameness as well—which is just not true. These early cells, in the preimplantation stage, are not in any biologically meaningful sense formed yet. Even at the eight-week stage when the embryo becomes a fetus, it has rudimentary form but lacks detail and cannot yet perform essential functions. These early stages are part of a continuity with the later stages, when functions will develop that are essential for human beings, but the early stages share little more than material with what we would see as an adult. It makes little sense biologically to give legal and constitutional rights to these early clusters of cells, as the Sanctity of Life bill would ask us to do.

Personhoodusa.com says that "Personhood is the cultural and legal recognition of the equal and unalienable rights of human beings,"[4] and they assert that embryos are human beings. Undoubtedly the supporters of these groups are sincere and do see embryos as just the earliest stages of us—which is true in some senses, of course. But whether embryos are "human beings" depends entirely on the definition of that term, and definitions are a matter of social convention on all sides. Most commonly, human beings have been defined in terms of man, woman, or child—not in terms of embryos. What the personhood proponents do not see, perhaps because they have not looked, is how those embryos appear under the microscope. Embryos are inert cells; they do not grow, or undergo any significant gene expression, or have any capacity to develop further until they are implanted into a gestating uterus.

Policy for the Implanted Embryo and Fetus to Twenty-Four Weeks

Now let us look at other developmental stages. At the point of implantation, the human embryo moves inside and out of sight. It becomes connected to the mother and, as mentioned earlier, is in effect dependent on her for nutrients and for disposing waste. The first eight weeks begin to involve gene expression, differentiation, and the emergence of form. As mentioned earlier, by the end of the eight weeks, typically the embryo will have all the major organ systems in place and has the outline of its form, but it is still lacking many essential functions. At this point, the embryo is called a fetus, and it begins to develop function as well as to refine the development of form.

The United States Supreme Court decided with *Roe v. Wade* that during the first two trimesters the embryo and fetus are part of the mother's body. This covers the period up until 24 weeks, which is defined in terms of time since the last menstrual period and means 22 weeks of actual gestation. The embryo/fetus is not independent, and according to U.S. law does not have any kind of legal personhood on its own during this period. Therefore, the mother may choose to continue a pregnancy or choose to end a pregnancy. The Court's reasoning has been discussed in depth by many, many commentators, and the main point here is that the Court declined to make new definitions or to attempt to do their own biological analysis of the fetus. But the Court did note that by the time of the third trimester of pregnancy, the fetus can be viable. That is a biological and medical recognition that the fetus under some rare circumstances is capable of living on its own after 24 weeks. The Court therefore left it to states to determine how they want to regulate pregnancies in the third trimester when the fetus might possibly, under some circumstance and at great expense, be viable.

The *Roe* decision therefore was consistent with the best biology of the day, and it remains so. Medical authorities universally agree that fetuses before twenty-four weeks are not viable and do not live outside the mother. Every day matters at that point. Only a very small number of fetuses that have been delivered as early as twenty-three weeks plus a few days rather than the full twenty-four weeks have survived; in each case, there were questions about the exact date of fertilization, and in every case there were serious medical problems or death very shortly after birth. If viability means "capable of living a little longer, even with severe defects," then some medical experts think that twenty-three-week fetuses may very occasionally have a chance of such life—but not before, because the organ systems are not working yet.

Fetal Pain

A recent political move by antiabortion groups centers on their attempts to define when pain occurs in fetuses. Here, the divergence of biological and metaphysical views comes directly into conflict through policy debates. The argument goes that fetuses can feel pain as early as twenty weeks, and therefore they should be protected from abortion, which clearly causes pain. The proponents of what has been called the fetal pain movement start with the social assumption that pain is bad, with which most everybody would agree. They add the assumption that inflicting pain intentionally is bad, with which most people would agree, although we know that many medical procedures do cause pain to achieve their desired results. Therefore, it is not possible to avoid inflicting pain entirely. Figuring out when it is acceptable and when it is not is another matter of biomedical ethics. Nonetheless, we can agree that intentionally inflicting avoidable pain is bad and should be avoided if possible.

But then we need to be able reliably to assess when we are, in fact, inflicting pain. This is a matter of biological or biomedical fact in the case of the fetus. We know that an embryo cannot feel pain because it does not have any neurological system or any pain receptors or processors. Neither does a fetus for most of its life, but the question arises at what point does it feel pain? We now have a movement insisting that its advocates have knowledge about that twentieth week, and they point to various pieces of what they consider to be evidence in favor of their claim, including the impression that the fetus moves in response to some stimuli.

Advocates of the fetal pain movement have succeeded in getting legislation passed, first in 2011 in Nebraska, then in a dozen more states within the two subsequent years. The debates surrounding whether fetuses feel pain have made it clear that the scientific facts actually matter very little to those pushing the legislation. Texas passed legislation in July 2013 in the form of House Bill 2, "Relating to the regulation of abortion procedures, providers, and facilities; providing penalties." When Governor Rick Perry signed the bill into law, Texas began to prohibit abortions at 20 weeks. More such bills are likely until the higher courts begin to rule on the legal issues. For now, the restriction on abortions at twenty weeks clearly violates the *Roe* decision that abortion is allowed for the first two trimesters, and the states are waiting for the inevitable conflicts between state and federal law to play out.

Assertions about when the fetus feels pain are an extension of metaphysical belief because they are not grounded in biologically verified facts. Let us look at this episode in a little more detail to see what is at stake and what are the social and policy implications. It matters who is considered to be an expert on the fetus's state of perception and in particular on the question whether a fetus can feel pain. Surely experts on such biological and neurological matters should be expert researchers in the field, those who have reviewed the literature and added to it, as the question about whether a fetus can physically feel pain is not a metaphysical but a biological one. Even those advocating the fetal pain restriction on abortions have adopted the strategic tactic of pointing to what they consider the biology, so it would seem that all parties agree that the biology is what matters for deciding the facts about fetal pain. This means researchers in neurodevelopment are the experts; they have been considering the question about fetal pain since the introduction of surgery on fetuses.

Fetal surgery raised important questions about standards of care and how to protect the fetus from pain as well as about the medical welfare of both mother and fetus. A useful review of the evidence available in 2005 came after Arkansas and Georgia had enacted restrictions on abortions, based on assertions about fetal pain. The article by a group of scientists, lawyers, and others in the *Journal of the American Medical Association* called for more study but also made clear that there was then no serious scientific evidence in favor of claims that fetuses feel pain at twenty weeks.[5] By 2007, an overview summary of the neurobiological evidence by Curtis Lowery and others showed clearly that the neural wiring is not in place yet for fetuses to feel pain at twenty weeks.[6] According to all the best evidence at hand, fetuses only gradually become able to sense the environment and to respond after week twenty-four, and many subsequent studies have continued to reinforce this scientific conclusion.[7]

What kinds of evidence do researchers rely on to assess fetal pain? Researchers have solidly established what neural wiring is necessary to allow sensations, the perception of those sensations, and the ability to interpret them as pain. They know this from studying fetuses that have been aborted naturally or clinically, and they also have studied animals. They have also examined the available images of fetuses in the womb, including the rare direct photographs such as those of Lennart Nilsson, and also images obtained by sonogram, ultrasound, and other technologies. All this evidence has provided a good scientific understanding of just when each of the neural parts responsible for pain and other sensations becomes active.

Ironically, fetal pain proponents rely on imaging technologies as well, but their interpretations of fetal images tend to diverge from the medical and scientific explanations. Whereas the scientific researchers can see clearly that the neural connections are not in place, antiabortion advocates may point to a picture of a fetus seeming to flinch. They claim that such examples allow us to "see" the pain. However, when the biological evidence has shown clearly that the fetus has no neural system in place yet, what the fetus is doing cannot by biological definition be a reaction to neural stimuli—it cannot be pain in any traditional biological sense. All the best biological evidence shows that the twenty-week-old fetus is not yet capable of feeling physical pain and to suggest otherwise is relying on metaphysical beliefs, not science.

A *New York Times* report shortly after the Texas legislative vote quoted the state policy director for the Right to Life Committee, Mary Spaulding Baluch, who is also a lawyer, as admitting that she is an advocate for ending abortion in general who "saw an opening" in pushing to limit late-stage abortions via the pain argument. The ban on abortions at twenty weeks is, in her opinion, "a step toward their ultimate goal": "Our mission is to restore legal protection to unborn life from the moment of conception," she said. "This is a marathon."[8] It is clear that the science really does not matter—only the political goal counts, which is the way politics often works. Those advocating restrictions on abortion at twenty weeks on the grounds that it causes fetal pain are relying on metaphysics and personal preference, not on science.

Conclusions

As we saw with the U.S. 2012 presidential election and related campaigns at the state and local levels, the embryo played a central role in many elections. As embryos have moved into the forefront of the public arena, diverging meanings and diverging ways of gaining knowledge about the world (called *epistemologies*) have been applied to them. The two divergent, often strident sets of claims in the debates often go something like the following:

> *Position 1:* I believe that an embryo is a person. I believe that the fetus feels pain at twenty weeks. Therefore, life is good, and abortion is always bad. This means that we must never do embryo research.
>
> *Position 2:* I know that an embryo is not fully formed and that it is very different from later developmental stages. I can see the differences, I can see the gradual emergence of form and function, and I know from neurobiological studies that fetuses cannot feel pain at twenty weeks. We can see fetal behaviors

> that do not conform to what we know about neural reactions during pain. More research, carried out responsibly, will help inform wise decisionmaking.

The first position relies on a metaphysical belief system, the second on an epistemological system grounded in scientific investigation, reason, and evidence. This divergence of meaning for embryos is not necessarily a distinction between science and ethics. Science does not tell us what is good or bad, but rather gives us the most accurate understanding of nature at any given time. Scientists care about ethics as well, and they want wise, responsible, and consistent policies, just as ethical citizens should want sound science.

We can best understand what embryos are by putting them under the microscope and looking at them carefully. Any policies that depend on a scientific understanding of embryos should be informed by or at the very least be consistent with the scientific evidence—anything less is to surrender to unwarranted beliefs. This means, in particular, that it does not make sense to argue that embryos are persons in the same sense as adults, children, or babies. It also means that because we have evidence that fetuses cannot feel pain at twenty weeks, we have no biological reason to restrict abortions at that twenty-week point.

If we take biological knowledge to be the appropriate basis for policy making on matters that include claims about science, we should legislate accordingly. Anything else, including some recent policies enacted at the state level, is tantamount to relying on metaphysics and claiming it as science.

Notes

Preface

1. Jane Maienschein, Manfred Laubichler, and Andrew Loettgers, "How Can History of Science Matter to Scientists?" *ISIS* 99 (2008): 341–349. AAAS, Scientia blog on history of science: http://membercentral.aaas.org/blogs/scientia. And a webinar on why the history of science matters: http://membercentral.aaas.org/multimedia/webinars/historys-role-science (accessed July 3, 2011).

1. Recurring Questions, Seeing and Believing

1. For a useful summary, see Theresa Phillips, "The Role of Methylation in Gene Expression," *Nature Education* 1 (2008): 1, www.nature.com/scitable/topicpage/the-role-of-methylation-in-gene-expression-1070.
2. Brian Hall is one who sees the neural crest as a definable germ layer: *The Neural Crest and Neural Crest Cells in Vertebrate Development and Evolution* (New York: Springer, 2009).
3. Sanctity of Human Life Act, H.R. 212, 112th Cong. (2011), http://thomas.loc.gov/cgi-bin/query/z?c112:H.R.212:.
4. National Conference of State Legislatures, "Stem Cell Research" (a summary of embryo-related bills), updated January 2008, at: www.ncsl.org/issues-research/health/embryonic-and-fetal-research-laws.aspx.

5. Dickey-Wicker Amendment, Pub. L. No. 104–134, 110 Stat. 1321–229 (1996). This is added to the official Balanced Budget Downpayment Act I of 1996, known as Public Law No 104-99, § 128, 110 Stat. 26, 34.
6. "Embryo, n. and adj.," OED Online, September 2013 (Oxford University Press), http://www.oed.com/view/Entry/61058.
7. For example, Scott Gilbert, Anna L. Tyler, and Emily Zacklin, *Bioethics and the New Embryology: Springboards for Debate* (New York: W. H. Freeman and Sinauer Associates, 2005). Gilbert and Robert are writing an updated look at the newest issues raised by embryo research.
8. Laurie Zoloth, "The Ethics of the Eighth Day: Jewish Bioethics and Research on Human Embryonic Stem Cells," in *The Human Embryonic Stem Cell Debate,* ed. Suzanne Holland, Karen Lebacq, and Laurie Zoloth, 95–79 (Cambridge, Mass.: MIT Press, 2001).
9. St. Augustine, "The Case of Abortive Conceptions," in *On the Holy Trinity; Doctrinal Treatises; Moral Treatises,* vol. 3 of *A Select Library of Nicene and Post-Nicene Fathers of the Christian Church,* ed. trans. Philip Schaff (ca. 420; Grand Rapids, Mich.: Wm. B. Eerdmans, 1890), www.ccel.org/ccel/schaff/npnf103.iv.ii.lxxxvii.html.
10. St. Thomas Aquinas, "Treatise on Man" (questions 75–102), in *Summa Theologica,* trans. Fathers of the English Dominican Province (1265–1274; New York: Benziger Bros., 1947), http://www.ccel.org/ccel/aquinas/summa.
11. Pius IX, *Apostolicae Sedis moderationi,* 1869.
12. Jane Maienschein, "What Determines Sex? A Study of Converging Approaches, 1880–1916," *Isis* 75 (1984): 457–480.

2. Hypothetical and Observed Embryos with Microscopes at Work

1. See Aristotle, *Physics* 2.3; and *Metaphysics* 5.2.
2. Andrea Falcon, "Aristotle on Causality," in *Stanford Encyclopedia of Philosophy* (Stanford University, 1997–), article published January 5, 2011, http://plato.stanford.edu/entries/Aristotle-causality/.
3. Aristotle, *Generation of Animals,* trans. A. L. Peck (Cambridge, Mass.: Harvard University Press, 1979).
4. For a wonderful view of what Aristotle may have observed, see "Aristotle's Lagoon," dir. Armand Marie Leroi, BBC, May 4, 2011, http://www.bbc.co.uk/programmes/p00gqlyy.

5. William Harvey, *Exercitationes de Generatione Animalium* (London: William Dugard for Octavian Pullwyn, 1651).
6. Shirley Roe, *Matter, Life, and Generation: Eighteenth Century Embryology and the Haller-Wolff* (Cambridge: Cambridge University Press, 1981).
7. For discussion of the issues in general as well as the particular views, see Roe, *Matter, Life, and Generation;* Peter J. Bowler. "Preformation and Pre-Existence in the Seventeenth Century: A Brief Analysis," *Journal of the History of Biology* 4 (1971): 221–244; William Coleman, *Biology in the Nineteenth Century: Problems of Form, Function and Transformation* (Cambridge: Cambridge University Press, 1978).
8. Howard Lenhoff and Sylvia G. Lenhoff, "Abraham Trembley and the Origins of Research and Regeneration in Animals," in *A History of Regeneration Research: Milestones in the Evolution of a Science,* ed. Charles E. Dinsmore, 47–66 (Cambridge: Cambridge University Press, 1991).
9. Karl Ernst von Baer, *De Ovi Mammalium et Hominis Genesi* (Lipsiae, 1827).
10. Karl Ernst von Baer, "Die Metamorphose des Eies Batrachier vor der Erscheinung des Embryo und Folgerunen aus ihr für die Theorie der Erzeugung," *(Müller's) Archiv für Anatomie, Physiologie, und wissenschaftliche Medizin* (1834): 481–508.
11. Charles Darwin, *On the Origin of Species by Means of Natural Selection; Or, the Preservation of Favoured Races in the Struggle for Life* (London: John Murray, 1859), 449.
12. Darwin, *Origin of Species,* 432–433.
13. Ibid., 433.
14. Ibid., 533.
15. Karl Ernst von Baer, *Über Entwickelungsgeschichte der Thiere,* trans. Thomas Henry Huxley as "On the Development of Animals, with Observations and Reflections," in *A Source Book in Animal Biology,* ed. Thomas S. Hall, 392–399 (Cambridge: Harvard University Press, 1951).
16. Robert J. Richards, *The Tragic Sense of Life: Ernst Haeckel and the Struggle over Evolutionary Thought* (Chicago: University of Chicago Press, 2008).
17. Darwin, *Origin of Species*, 490.
18. Karen L. Wellner, "From Fertilization to Birth: Representing Development in High School Biology Textbooks" (master's thesis, Arizona State University, 2010).

19. Oscar Hertwig, *Die Zelle und die Gewebe: Grundzüge der Allgemeinen Anatomie und Physiologie* (Jena: Fischer, 1893).
20. Edmund Beecher Wilson, *The Atlas of Fertilization and Karyokinesis of the Ovum* (New York: Macmillan for Columbia University Press, 1895); Edmund Beecher Wilson, *The Cell in Development and Inheritance* (New York: Macmillan, 1896).
21. William Thompson Sedgwick and Edmund Beecher Wilson, *An Introduction to General Biology,* (New York, Henry Holt, 1887).
22. Wilson noted that a biologist studying such things had to have a very patient wife, or one who helped him. Wilson's wife Anne Maynard Kidder, was apparently both, and their daughter Nancy became a fine musician who also helped out in the laboratory on occasion.
23. Wilson, *Atlas of Fertilization.*
24. Wilson, *The Cell in Development and Inheritance.*
25. Wilson, *The Cell in Development and Inheritance;* 2nd. ed. (New York: Macmillan, 1900); *The Cell in Development and Heredity,* 3rd ed. (New York: Macmillan, 1925).
26. "MBL President and Director Gary Borisy to Receive Award for 'Far-Reaching Contributions to Cell Biology,'" press release, Marine Biological Laboratory, May 16, 2011, http://hermes.mbl.edu/news/press_releases/2011/2011_pr_05_16.html.
27. *Biological Lectures Presented at the Marine Biological Laboratory in Woods Hole,* 1890–1893, available in the Biodiversity Heritage Library at www.biodiversitylibrary.org/bibliography/4963.
28. "The Nobel Prize in Physiology or Medicine 2002: Sydney Brenner, H. Robert Horvitz, John E. Sulston," press release, NobelPrize.org, October 7, 2002, www.nobelprize.org/nobel_prizes/medicine/laureates/2002/press.html.
29. Nick Hopwood, "Producing Development: The Anatomy of Human Embryos and the Norms of Wilhelm His," *Bulletin of the History of Medicine* 74 (2000): 29–79; Nick Hopwood, *Embryos in Wax: Models from the Ziegler Studio* (Cambridge: Whipple Museum of the History of Science, 2002).
30. Frederic L. Holmes, "The Old Martyr of Science: The Frog in Experimental Physiology," *Journal of the History of Biology* 26 (1993): 311–328.
31. Ronan O'Rahilly and Fabiola Müller, *Developmental States in Human Embryos,* Publication 637 (Washington, D.C.: Carnegie Institute of

Washington, 1987); Jane Maienschein, Marie Glitz, and Garland E. Allen, eds., *The Department of Embryology,* vol. 5 of *Centennial History of the Carnegie Institution of Washington* (Cambridge: Cambridge University Press, 2004).

32. Wilhelm His, *Unsere Körperform und das physiologische Problem ihrer Entstehung: Briefe an einen Befreundeten Naturforscher* (Leipzig: F. C. W. Vogel, 1874).

3. Experimental Embryos in the Laboratory

1. Garland E. Allen, *Life Science in the Twentieth Century* (Cambridge: Cambridge University Press, 1975).
2. Jane Maienschein, Ronald Rainger, and Keith R. Benson, "Morphology and Modern Biology: Were the Americans in Revolt?" *Journal of History of Biology* 14 (1981): 83–87. Also see the articles of the special section on "American Morphology at the Turn of the Century."
3. Wilehlm Roux, "The Problems, Methods, and Scope of Developmental Mechanics," in *Biological Lectures Delivered at the Marine Biological Laboratory of Woods Hole 1894,* trans. William Morton Wheeler, 149–190 (Boston: Ginn & Co., 1895).
4. Viktor Hamburger, "Wilhelm Roux: Visionary with a Blind Spot," *Journal of the History of Biology* 30 (1997): 229–238.
5. Wilhelm Roux, "Beiträge zur Entwickelungsmechanik des Embryo, No. 5. Über die künstliche Hervorbringung halber Embryonen durch Zerstörung einer der beiden ersten Furchungskugeln, sowie über die Nachentwickelung (Postgeneration) der fehlenden Körperhälfte," *Virchow's Archiv für Pathologisches Anatomie und Physiologie und klinische Medizin* 114 (1888): 113–153; reprinted in *Foundations of Experimental Embryology,* trans. Benjamin Willier and Jane M. Oppenheimer, 1–37 (New York: Hafner, 1964).
6. Recent work confirms what has long been suspected, namely, that even different cells are genetically distinct because of mutations and other imperfections of cell division. The result is a genetic mosaic, as discussed in James R. Lupski, "Genome Mosaicism—One Human, Multiple Genomes," *Science* 341 (2013): 358–359.
7. Hans Driesch, "Enwicklungsmechanische Studien. I. Der Werth der beiden ersten Furchungszellen in der Echinodermentwicklung. Experimentelle Erzeugen von Theil- und Doppelbildung," *Zeitschrift für*

wissenschaftliche Zoologie 53 (1891): 160–178; reprinted in *Foundations of Experimental Embryology*, trans. Benjamin Willier and Jane M. Oppenheimer, 38–50 (New York: Hafner, 1964).

8. Manfred D. Laubichler and Eric H. Davidson, "Boveri's Long Experiment: Sea Urchin Merogones and the Establishment of the Role of Nuclear Chromosomes in Development," *Developmental Biology* 314 (2008): 1–11.
9. Edmund Beecher Wilson, *The Cell in Development and Inheritance* (New York: Macmillan, 1896), 330.
10. Philip Pauly, *Controlling Life: Jacques Loeb and the Engineering Ideal in Biology* (New York: Oxford University Press, 1987).
11. Kenneth R. Manning, *Black Apollo of Science: The Life of Ernest Everett Just* (New York: Oxford University Press, 1983).
12. Thomas Hunt Morgan, *Regeneration* (New York: Macmillan, 1901).
13. Mary Evelyn Sunderland, "Regeneration: Thomas Hunt Morgan's Window into Development," *Journal of the History of Biology* 43 (2010): 325–361.
14. Garland E. Allen, *Thomas Hunt Morgan: The Man and His Science* (Princeton, N.J.: Princeton University Press, 1978). See also Jane Maienschein, *Transforming Traditions in American Biology, 1880–1915* (Baltimore: Johns Hopkins University Press, 1991).
15. Sunderland, "Regeneration"; Morgan, *Regeneration.*
16. My used copy of Morgan's *Regeneration* cost $35 years ago; today, copies are difficult to find and cost hundreds of dollars. The book is also available through print on demand, reflecting the renewed interest in his work.
17. Thomas Hunt Morgan, "Regeneration and Liability to Injury," *Zoological Bulletin* 1 (1898): 287–300, http://www.jstor.org/stable/1535478.
18. Ibid., 293, 299.
19. Thomas Hunt Morgan, "Further Experiments on the Regeneration of Tissue Composed of Parts of Two Species," *Biological Bulletin* 2 (1900): 111–119.
20. Ibid., 115.
21. Thomas Hunt Morgan, "The Dynamic Factor in Regeneration," *Biological Bulletin* 16 (1909): 265–276.
22. Ibid., 256.
23. Thomas Hunt Morgan, *Embryology and Genetics* (New York: Columbia, 1934).

24. Ibid., 169.
25. Ibid., 234.
26. Hans Spemann, *Embryonic Development and Induction* (New Haven, Conn.: Yale University Press, 1938).
27. Johannes Holftreter, "Reminiscences on the Life and Work of Johannes Holtfreter," in *A Conceptual History of Modern Embryology*, ed. Scott F. Gilbert, 109–127 (Baltimore: Johns Hopkins University Press, 1991).
28. Jane Maienschein, "Regenerative Medicine's Historical Roots in Regeneration, Transplantation, and Translation," *Developmental Biology* 358 (2011): 278–284.
29. Ross Granville Harrison, "The Outgrowth of the Nerve Fiber as a Mode of Protoplasmic Outgrowth," *Journal of Experimental Zoology* 9 (1910): 787–846. This article was the full report on a technique first published in 1907.
30. Alexis Carrel, "On the Permanent Life of Tissues Outside of the Organism," *Journal of Experimental Medicine* 15 (1912): 516–528; Alexis Carrel and Charles A. Lindbergh, *The Culture of Organs* (New York: Paul B. Hoeber, 1938).
31. Hannah Landecker, *Culturing Life: How Cells Became Technologies* (Cambridge, Mass.: Harvard University Press, 2007).
32. Rebecca Skloot, *The Immortal Life of Henrietta Lacks* (New York: Crown Books, 2010).
33. Ross Granville Harrison, "Wound Healing and Reconstitution of the Central Nervous System of the Amphibian Embryo after Removal of Parts of the Neural Plate," *Journal of Experimental Zoology* 106 (1974): 27–84.
34. Ibid., 47.
35. John Spangler Nicholas, "Ross Granville Harrison 1870–1959: Biographical Memoirs," *National Academy of Science of the United States* (1961): 130–162; Samuel R. Detwiler, *Neuroembryology* (New York: Macmillan, 1936).
36. Viktor Hamburger, "The History of the Discovery of the Nerve Growth Factor," *Journal of Neurobiology* 24 (1993): 893–897.

4. Inherited, Evolved, and Computed Embryos

1. James D. Watson and Francis H. C. Crick, "A Structure for Deoxyribose Nucleic Acid," *Nature* 171 (1953): 737–738.

2. Dorothy Nelkin and Susan Lindee, *The DNA Mystique: The Gene as a Cultural Icon* (New York: W. H. Freeman, 1995).
3. The Ross Granville Harrison Papers in the archives at Yale University's Sterling Library make clear Harrison's views on genetics, which he repeated in letters and discussions with colleagues and students.
4. Jane Oppenheimer, personal discussions during the late 1970s as I was completing my dissertation on Ross Harrison's work. Oppenheimer was an embryologist, whose important contributions to the history of embryology included exploring the move to genetics from a focus on issues of morphogenesis.
5. Jan Sapp, *Beyond the Gene: Cytoplasmic Inheritance and the Struggle for Authority in Genetics* (New York: Oxford University Press, 1987).
6. Thomas Hunt Morgan, "Chromosomes and Heredity," *American Naturalist* 44 (1910): 449–496.
7. Frank Rattray Lillie, "The Theory of the Free-Martin," *Science* 63 (1916): 611–615.
8. Evelyn Fox Keller, *The Mirage of a Space between Nature and Nurture* (Durham, N.C.: Duke University Press, 2010).
9. James D. Watson and F. H. C. Crick, "A Structure for Deoxyribose Nucleic Acid," *Nature* 171 (1953): 737–738.
10. See, for example, Lily Kay, *The Molecular Vision of Life: Caltech, the Rockefeller Foundation, and the Rise of the New Biology* (New York: Oxford University Press, 1996); Michel Morange, *A History of Molecular Biology* (Cambridge, Mass.: Harvard University Press, 2000).
11. For information on the research leading to the Nobel Prize, for knockout techniques, see "The Nobel Prize in Physiology or Medicine 2007: Mario R. Capecchi, Sir Martin J. Evans, Oliver Smithies," press release, NobelPrize.org, October 8, 2007, www.nobelprize.org/nobel_prizes/medicine/laureates/2007/press.html.
12. Robert Briggs and Thomas J. King, "Transplantation of Living Nuclei from Blastula Cells into Enucleated Frogs' Eggs," *Proceedings of the National Academy of Sciences of the United States of America* (1952) 38: 455–463.
13. Cited by Nicholas Wade, "Cloning and Stem Cell Work Earns Nobel," *New York Times,* October 8, 2012, www.nytimes.com/2012/10/09/health/research/cloning-and-stem-cell-discoveries-earn-nobel-prize-in-medicine.html.

14. John B. Gurdon, "The Developmental Capacity of Nuclei Taken from Intestinal Epithelium Cells of Feeding Tadpoles," *Journal of Embryology and Experimental Morphology* 34 (1962): 93–112.
15. Roy J. Britten and Eric Davidson, "Gene Regulation for Higher Cells: A Theory," *Science* 165 (1969): 349–357.
16. Ibid., 349.
17. Rachel Ankeny, "Historiographic Reflections on Model Organisms: Or How the Mureaucracy May Be Limiting Our Understanding of Contemporary Genetics and Genomics," *History and Philosophy of the Life Sciences* 32 (2010): 91–104.
18. See Davidson's website for an overview, at www.its.caltech.edu/~davidson/; and Eric H. Davidson, *Gene Activity in Early Development* (New York: Academic Press, 1977); Eric H. Davidson, *The Regulatory Genome: Gene Regulatory Networks in Development and Evolution* (New York: Academic Press, 2006).
19. Nick Hopwood, *Embryos in Wax: Models from the Ziegler Studio* (Cambridge: Whipple Museum of the History of Science, 2002).
20. Rachel Fink, *A Dozen Eggs: Time-Lapse Microscopy of Normal Development* (Sunderland, Mass.: Sinauer Associates, 1991).
21. See, for example, Philipp J. Keller, "Imaging Morphogenesis: Technological Advances and Biological Insights," *Science* 340 (2013): 1184. That morphogenesis and imaging are central topics for developmental biology is made clear in a special section on "Morphogenesis" in *Science* 340 (2013): 1183–1194. On the challenges that come with such work, see Vivien Marx, "The Big Challenges of Big Data," *Nature* 498 (2013): 255–260.

5. The Visible Human Embryo

1. Martin Johnson, "Robert Edwards: Nobel Laureate in Physiology or Medicine," in *Les prix Nobel 2010,* ed. Karl Grandin, 237–256 (Stockholm: Nobel Foundation, 2010); "The Nobel Prize in Physiology or Medicine 2010: Robert G. Edwards," press release, Nobelprize.org, October 10, 2010, www.nobelprize.org/nobel_prizes/medicine/laureates/2010/press.html.
2. Johnson, "Robert Edwards," 246.
3. Robert G. Edwards and Patrick Steptoe, *A Matter of Life: The Story of a Medical Breakthrough* (London: Hutchinson, 1980).

4. For discussion, see articles in Jane Maienschein, Marie Glitz, and Garland E. Allen, eds., *Centennial History of the Carnegie Institution of Washington Department of Embryology* (Cambridge: Cambridge University Press, 2004).
5. Lennart Nilsson, *A Child Is Born: The Drama of Life before Birth in Unprecedented Photographs: A Practical Guide for the Expectant Mother,* text by Axel Ingelman-Sundberg and Claes Wirse'n, trans. Britt Wirse'n, Claes Wirse'n, and Annabelle MacMillan (New York: Delacorte Press, 1967), originally published in Swedish as *Ett barn blir till,* 1965.
6. Lennart Nilsson, *A Child Is Born,* 4th ed., text by Lars Hamberger, trans. Linda Schenck (New York: Delacourte Press, 2003).
7. *The Miracle of Life,* directed by Mikael Agaton, photography by Lennart Nilsson (Sweden 1982; U.S. 1983; Boston: WGBH, 2006), DVD.
8. See Scott Gilbert's website http://www.swarthmore.edu/academics/biology/faculty-and-staff/scott-gilbert/publications.xml for examples of these speeches and writings.
9. "Report of the Committee of Inquiry into Human Fertilisation and Embryology," submitted to the Parliament from Dame Mary Warnock (London: Her Majesty's Stationery Office, 1984), iv (available at www.hfea.gov.uk/docs/Warnock_Report_of_the_Committee_of_Inquiry_into_Human_Fertilisation_and_Embryology_1984.pdf).
10. Ibid., para 1.4.
11. Ibid., para. 11.17.
12. Ibid., para. 11:30.
13. Gregory Pincus, *The Control of Fertility* (New York: Academic Press, 1968).
14. Nicola Biesel, *Imperiled Innocents: Anthony Comstock and Family Reproduction in Victorian America* (Princeton, N.J.: Princeton University Press, 1998).
15. *Roe v. Wade* 410 U.S. 113 (1973).
16. Ibid.
17. Ibid.
18. Etienne-Emile Baulieu with Mort Rosenblum, *The "Abortion Pill" RU-486: A Woman's Choice* (New York: Simon and Schuster, 1991), 15.
19. Pam Belluck, "Abortion Qualms on Morning-After Pill May Be Unfounded [No Abortion Role Seen for Morning-After Pill]," *New York Times,* June 5, 2012, www.nytimes.com/2012/06/06/health/research

/morning-after-pills-dont-block-implantation-science-suggests.html; Pam Belluck, "Drug's Nickname May Have Aided Politicization," June 5, 2012, www.nytimes.com/2012/06/06/health/research/medications-nickname-may-have-helped-in-its-politicization.html.

20. See http://www.csmonitor.com/USA/Politics/2012/0516/Mitt-Romney-raising-money-at-home-of-morning-after-pill-exec.
21. Mayo Clinic Staff, "Morning-After Pill: Definition," MayoClinic.com, June 28, 2013, www.mayoclinic.com/health/morning-after pill/MY01190/DSECTION=why-its-done.
22. See http://embryo.asu.edu/pages/davis-v-davis-brief-1992.
23. See, for example, http://caselaw.findlaw.com/tx-court-of-appeals/1048566.html.
24. See, for example, http://embryo.asu.edu/pages/jeter-v-mayo-clinic-arizona-brief-2005 and http://caselaw.findlaw.com/az-court-of-appeals/1351932.html.
25. Susan Young, "Mississippi to Vote on 'Personhood,'" *Nature* 479 (2011): 13–14.
26. Lynn Morgan, "The Rise and Demise of a Collection of Human Fetuses at Mount Holyoke College," *Perspectives in Biology and Medicine* 49 (2006): 435–451.
27. June Goodfield, *Playing God: Genetic Engineering and the Manipulation of Life* (New York: Random House, 1977); Sheldon Krimsky, *Genetic Alchemy: The Social History of the Recombinant DNA Controversy* (Cambridge, Mass.: MIT Press, 1982).
28. Maxine Singer, "The Recombinant DNA Debate," *Science* 196 (1977): 127.
29. For a discussion of the eugenics movement and its underpinning science and implications, see Diane Paul, *Controlling Human Heredity: 1865 to the Present* (Atlantic Highlands, N.J.: Humanities Press, 1995); Daniel Kevles, *In the Name of Eugenics: Genetics and the Uses of Human Heredity* (Cambridge, Mass.: Harvard University Press, 1995).
30. Mark H. Haller, *Eugenics: Hereditarian Attitudes in American Thought* (New Brunswick, N.J.: Rutgers University Press, 1963), 141.
31. See, for example, Garland E. Allen, "Eugenics and Modern Biology: Critiques of Eugenics, 1910–1945," *Annals of Human Genetics* 75 (2011): 314–325.
32. For an excellent critique of the problems created by what has often been presented as a dichotomy of nature versus nurture, see Evelyn

Fox Keller, *The Mirage of a Space between Nature and Nurture* (Durham, N.C.: Duke University Press, 2010).

6. The Idea of Engineered and Constructed Embryos

1. Jacques Loeb, *The Mechanistic Conception of Life,* ed. Donald Fleming (Cambridge, Mass.: Harvard University Press, 1964).
2. Philip Pauly, *Controlling Life: Jacques Loeb and the Engineering Ideal in Biology* (New York: Oxford University Press, 1987).
3. Robert E. Kohler, *Partners in Science: Foundations and Natural Scientists, 1900–1945* (Chicago: University of Chicago Press, 1991).
4. Jacques Loeb, "On the Nature of the Process of Fertilization and the Artificial Production of Normal Larvae (Plutei) from the Unfertilized Eggs of Sea Urchins," *American Journal of Physiology* 3 (1899): 135–138.
5. For more discussion, see Pauly, *Controlling Life.*
6. Such episodes were one of the motivations for the MBL later creating the Logan Science Journalism Program, which brings journalists to research laboratories to learn about science and gives scientists the opportunity to learn how to communicate about their work.
7. Jacques Loeb, "The Mechanistic Conception of Life," *Popular Science Monthly* 80 (1912): 5–21; and in *Mechanistic Conception of Life,* 3–31.
8. Stéphan Leduc, *Théorie physico-chimique de la vie et generations spontanées* (Paris: A Poinat, 1910). For more on Leduc, see Robert H. Carlson, *Biology Is Technology: The Promise, Peril, and New Business of Engineering Life* (Cambridge, Mass.: Harvard University Press, 2011), 35. Or for Leduc in historical perspective see Evelyn Fox Keller, *Making Sense of Life. Explaining biological Development with Models, Metaphors, and Machines* (Cambridge: Harvard University Press, 2002).
9. Edmund Vincent Cowdry, *General Cytology: A Textbook of Cellular Structure and Function for Students of Biology and Medicine* (Chicago: Chicago University Press, 1924); *Special Cytology: The Form and Function of the Cell in Health and Disease; A Textbook for Students of Biology and Medicine,* 2 vols. (New York: P. B. Hoeber, 1928).
10. Jean Brachet and Alfred E. Mirsky, eds., *The Cell: Biochemistry, Physiology, Morphology,* 6 vols. (New York: Academic Press, 1961).

11. William Bechtel, *Discovering Cell Mechanisms: The Creation of Modern Cell Biology* (Cambridge: Cambridge University Press, 2006).
12. John Tyler Bonner, *Morphogenesis: An Essay on Development* (Princeton, N.J.: Princeton University Press, 1952); Mary Sunderland, "Morphogenesis, Dictyostelium, and the Search for Shared Developmental Processes," *Studies of the History and Philosophy of the Biomedical Sciences* 42 (2011):508–517.
13. Leonard Hayflick and Paul S. Moorhead, "The Serial Cultivation of Human Diploid Cell Strains," *Experimental Cell Research* 25 (1961): 585–621; Leonard Hayflick, "The Limited in Vitro Lifetime of Human Diploid Cell Strains," *Experimental Cell Research* 37 (1965): 614–636.
14. Lijing Jiang, "Degeneration in Miniature: History of Cell Death and Aging Research in the Twentieth Century" (Ph.D. diss., Arizona State University, 2013).
15. Meredith Wadman, "Medical Research: Cell Division," *Nature* 498 (2013): 422–426, www.nature.com/news/medical-research-cell-division-1.13273.
16. Ibid.
17. "The Nobel Prize in Physiology or Medicine 2009: Elizabeth H. Blackburn, Carol W. Greider, Jack W. Szostak," NobelPrize.org, October 2009, www.nobelprize.org/nobel_prizes/medicine/laureates/2009/.
18. Eugene H. Bell, Paul Ehrlich, David Buttle, and Takaka Nakatsuji, "Living Tissue Formed in Vitro and Accepted as Skin-Equivalent Tissue of Full Thickness," *Science* 211 (1981): 1052–1054.
19. Ibid., 1054.
20. From the home page of the Eugene Bell Center for Regenerative Biology and Tissue Engineering at the Marine Biological Laboratory at Woods Hole, Mass., http://www.mbl.edu/bell/.
21. C. D. Ford, J. L Hamerton, D. W. H. Bernes, J. F. Loutit. "Cytological Identification of Radiation-Chimeras," *Nature* 177 (1956): 452–454.
22. Beatrice Mintz. "Formation of Genetically Mosaic Mouse Embryos," *American Zoologist* 2 (1962): 432 [abstract]. See also Beatrice Mintz, "Experimental Recombination of Cells in the Developing Mouse Egg: Normal and Lethal Mutant Genotypes," *American Zoologist* 2 (1962): 541–542; Beatrice Mintz. "Experimental Study of the Developing Mammalian Egg: Removal of the Zona Pellucida," *Science* 138 (1962): 594–595.

23. For further discussion of this line of research, see Nicole Le Douarin and Anne McClaren, eds., *Chimeras in Developmental Biology* (London: Academic Press, 1984).
24. "The Nobel Prize in Physiology or Medicine 2007: Mario R. Capecchi, Sir Martin J. Evans, Oliver Smithies," NobelPrize.org, December 2007, www.nobelprize.org/nobel_prizes/medicine/laureates/2007/.
25. Jason Scott Robert, "The Science and Ethics of Making Part-human Animals in Stem Cell Biology," *FASEB Journal* 20 (2006): 838–845.
26. Human Fertilisation and Embryology Authority, "Review of Hybrids & Chimeras," updated January 12, 2012, www.hfea.gov.uk/519.html.
27. See, for example, "Chimeric Embryos May Soon Get Their Day in the Sun," *Science* 340 (2013): 1509, 1560.
28. See, for example, work by T. Hata, S. Uemoto, and E. Kobayashi, "Transplantable Liver Production Plan: 'Yamaton'-Liver Project, Japan," *Organogenesis* 9, no. 4 (2013): DOI 10.4161/org.25760.
29. Jason Scott Robert and Françoise Baylis, "Crossing Species Boundaries," *American Journal of Bioethics* 3 (2003): 1–13.
30. I. Wilmut, A. E. Schnieke, J. McWhir, A. J. Kind, and K. H. Campbell, "Viable Offspring Derived from Fetal and Adult Mammalian Cells," *Nature* 385 (1997): 810–813.
31. For discussion of these cloning discoveries, see Ian Wilmut, Keith Campbell, and Colin Tudge, *The Second Creation: The Age of Biological Control by the Scientists Who Cloned Dolly* (London: Headline Press: 2000); Anne McLaren, "Pathways of Discovery. Cloning: Pathways to a Pluripotent Future," *Science* 288 (2000): 1775–1780. See also John Gurdon and Alan Colman's reflections on the earlier research in their excellent review "The Future of Cloning," *Nature* 402 (1999): 743–746.
32. Robin McKie, "Scientists Clone Adult Sheep," *The Observer,* February 22, 1997, www.theguardian.com/uk/1997/feb/23/robinmckie.theobserver.
33. Gina Kolata, "Scientist Reports First Cloning Ever of Adult Mammal," *New York Times,* February 23, 1997, http://www.nytimes.com/1997/02/23/us/scientist-reports-first-cloning-ever-of-adult-mammal.html.
34. The way the *New York Times* highlighted the exotic implications of the Dolly story seems questionable, but Kolata explained at a recent conference that for her science reporting is just another form of entertainment,

and she feels no responsibility for the impact a story has. Gina Kolata, keynote presentation at "The Rightful Place of Science?" Conference at Arizona State University, sponsored by the Consortium for Science, Policy, and Outcomes, May 17, 2010; Sean Hays, "Keynote: Apparently, Vannevar Bush Was Right All Along," The Rightful Place of Science? Blog Archive, May 17, 2010, http://cspo.events.asu.edu/?p=365.

35. Silver said he had been writing a book that addressed cloning at the time, yet he was completely astonished by the report. Lee M. Silver, *Remaking Eden: How Genetic Engineering and Cloning Will Transform the American Family* (New York: Avon Books, 1997).
36. See discussion of all this work in Wilmut, Campbell, and Tudge, *The Second Creation.*
37. See, for example, Sarah Franklin, *Dolly Mixtures: The Remaking of Genealogy* (Durham, N.C.: Duke University Press, 2007).
38. Woo Suk Hwang, Young June Ryu, Jong Hyuk Park, Eul Soon Park, Eu Gene Lee, Ja Min Koo, Hyun Yong Jeon, Byeong Chun Lee, Sung Keun Kang, Sun Jong Kim, et al., "Evidence of a Pluripotent Human Embryonic Stem Cell Line Derived from a Cloned Blastocyst" [retracted], *Science* 303 (2004): 1669–1674; Woo Suk Hwang, Sung Il Roh, Byeong Chun Lee, Sung Keun Kang, Dae Kee Kwon, Sue Kim, Sun Jong Kim, Sun Woo Park, Hee Sun Kwon, Chang Kyu Lee, et al., "Patient-Specific Embryonic Stem Cells Derived from Human SCNT Blastocysts" [retracted], *Science* 308 (2005): 1777–1783.
39. David Magnus and Mildred K. Cho, "Issues in Oocyte Donation for Stem Cell Research," *Science* 308 (2005): 1747–1748.
40. For the U.N. resolution, see Resolution Adopted by the General Assembly on 8 March 2005: 59/280, United Nations Declaration on Human Cloning, 59th session, A/RES/59/280, UN.org, http://daccess-dds-ny.un.org/doc/UNDOC/GEN/N04/493/06/PDF/N0449306.pdf.
41. "General Assembly Adopts United Nations Declaration on Human Cloning by Vote of 84-34-37," press release, UN.org, August 3, 2005, www.un.org/News/Press/docs/2005/ga10333.doc.htm.
42. Byunghun Hyun, "A Contextual Understanding of the Definition of Science in South Korea" (master's thesis, Arizona State University, 2011).
43. Masahito Tachibana, Paula Amato, Michelle Sparman, Nuria Marti Gutierrez, Rebecca Tippner-Hedges, Hong Ma, Eunju Kang, Alimujiang Fulati, Hyo-Sang Lee, Hathaitip Sritanaudomchai, et al., "Human

Embryonic Stem Cells Derived by Somatic Cell Nuclear Transfer," *Cell* 153 (2013): 1228–1238.

44. Gretchen Vogel, "Human Stem Cells from Cloning, Finally," *Science* 340 (2013): 795.
45. David Cyranoski, "Human Stem Cells Created by Cloning," *Nature* 497 (2013): 295–296.
46. "Endangered Species Cloned," *BBC News,* October 8, 2000, http://news.bbc.co.uk/2/hi/science/nature/962159.stm; "Endangered Animal Clone Dies," *BBC News,* January 12, 2001, http://news.bbc.co.uk/2/hi/science/nature/1113719.stm.
47. Ed Yong, "Resurrecting the Extinct Frog with a Stomach for a Womb," *National Geographic,* March 15, 2003, http://phenomena.nationalgeographic.com/2013/03/15/resurrecting-the-extinct-frog-with-a-stomach-for-a-womb/; Brian Howard and Christine Dell'Amore, "Back from the Dead?" *National Geographic,* March 16, 2013, http://news.nationalgeographic.com/news/2013/03/pictures/130316-gastric-brooding-frog-animals-weird-science-extinction-tedx/.
48. Carl Zimmer, "Bringing Them Back to Life," *National Geographic,* April 2013, http://ngm.nationalgeographic.com/2013/04/125-species-revival/zimmer-text.
49. Ricki Lewis, "A Stem Cell Legacy: Leroy Stevens," *The Scientist* 14 (2000): 19.
50. Leroy Stevens, "The Development of Transplantable Teratocarcinomas from Intratesticular Grafts of Pre- and Postimplantation Mouse Embryos," *Developmental Biology* 21 (1970): 364–382.
51. Beatrice Mintz and Karl Illmensee, "Normal Genetically Mosaic Mice Produced from Malignant Teratocarcinoma Cells," *Proceedings of the National Academy of Science of the United States of America* 72 (1975): 3585–3589.
52. Ricki Lewis, "A Stem Cell Legacy: Leroy Stevens," *The Scientist,* March 6, 2000, www.the-scientist.com/?articles.view/articleNo/12717/title/A-Stem-Cell-Legacy–Leroy-Stevens/.
53. Martin J. Evans and Matthew H. Kaufman, "Establishment in Culture of Pluripotential Cells from Mouse Embryos," *Nature* 292 (1981): 154–156.
54. Gail Martin, "Isolation of a Pluripotent Cell Line from Early Mouse Embryos Cultured in Medium Conditioned by Teratocarcinoma Stem

Cells," *Proceedings of the National Academy of Sciences of the United States of America* 78 (1981): 7634–7638.

55. See Jane Maienschein, *Whose View of Life? Embryos, Cloning, and Stem Cells* (Cambridge, Mass.: Harvard University Press, 2003), for more discussion of stem cell science in its social context.
56. Jeff Nisker, Françoise Baylis, Isabel Karpin, Carolyn McLeod, and Roxanne Mykitiuk, eds., *The Healthy Embryo: Social, Biomedical, Legal and Philosophical Perspectives* (Cambridge: Cambridge University Press, 2009).
57. Diane B. Paul, "The History of Newborn Phenylketonuria Screening in the U.S.," in *Promoting Safe and Effective Genetic Testing in the United States: Final Report of the Task Force on Genetic Testing,* ed Neil A. Holtzman and Michael S. Watson, 1–13 (Baltimore: Johns Hopkins University Press, 1998).
58. Edmund Beecher Wilson, *The Cell in Development and Inheritance* (New York: Macmillan, 1896), 330.
59. John Burris, Robert Cook-Deegan, and Bruce Alberts, "The Human Genome Project after a Decade: Policy Issues," *Nature Genetics* 20 (1998): 333–335.
60. See "Getting Started Is Simple," 23andMe.com, 2007–2013, www.23andme.com/howitworks/.
61. Gary Marchant, Arizona State University professor, personal discussions and informal presentations.

7. Constructing Embryos for Society, Stem Cells in Action

1. Vannevar Bush, *Science—The Endless Frontier: A Report to the President* (Washington, D.C.: U.S. Government Printing Office, 1945), www.nsf.gov/od/lpa/nsf50/vbush1945.htm.
2. Ibid. For a summary of the National Science Foundation's history, see the document by former NSF historian George T. Mazuzan, "The National Science Foundation: A Brief History," NSF 88-16, July 15, 1994, www.nsf.gov/about/history/nsf50/nsf8816.jsp. Also see Gregg Pascal Zachary's excellent biography of Bush: *Endless Frontier: Vannevar Bush, Engineer of the American Century* (New York: Free Press, 1997).
3. Bush, *Science—The Endless Frontier.*

4. National Science Foundation Act of 1950, Pub. L. No. 81-507, Stat. 149 (1950). A set of goals the NSF in 2013 still regards as its core mission: "NSF at a Glance," www.nsf.gov/about/glance.jsp.
5. Gerald E. Geison, *The Private Science of Louis Pasteur* (Princeton, N.J.: Princeton University Press, 1995).
6. Donald E. Stokes, *Pasteur's Quadrant: Basic Science and Technological Innovation* (Washington, D.C.: Brookings Institution Press, 1997).
7. Arizona State is my university, and the reader might suspect that I am biased in pointing to our president Michael Crow as a leader in this particular movement through use-inspired research. Yet his articles in prominent peer-reviewed journals and his selection as a leading speaker at high-profile conferences provides evidence for his leadership.
8. Michael Crow's vision for Arizona State University is articulated in a number of places, including here, through the president's office at ASU: "A New American University," http://newamericanuniversity.asu.edu/.
9. See "About NIH," National Institutes of Health, updated May 15, 2013, www.nih.gov/about/mission.htm.
10. Michael M. Crow, "Time to Rethink the NIH," *Nature* 471 (2011): 569–571.
11. Ibid.
12. Ibid.
13. The National Conference of State Legislatures provides a summary of state level funding and regulation, including of stem cell research at www.ncsl.org/default.aspx?tabid=14413.
14. James A. Thomson, Joseph Itskovitz-Eldor, Sander S. Shapiro, Michelle A. Waknitz, Jennifer J. Swiergiel, Vivienne S. Marshall, and Jeffrey M. Jones, "Embryonic Stem Cells Lines Derived from Human Blastocysts," *Science* 282 (1998): 1145–1147; Michael J. Shamblott, Joyce Axelman, Shunping Wang, Elizabeth M. Bugg, John W. Littlefield, Peter J. Donovan, Paul D. Blumenthal, George R. Huggins, and John D. Gearhart, "Derivation of Pluripotent Stem Cells from Cultured Human Primordial Germ Cells," *Proceedings of the National Academy of Sciences of the United States of America* 95 (1998): 13726–13731; J. D. Gearhart, "New Potential for Human Embryonic Germ Cells," *Science* 282 (1998): 1061–1062.
15. Beatrice Mintz, "Experimental Recombination of Cells in the Developing Mouse Egg: Normal and Lethal Mutant Genotypes," *American*

Zoologist 2 (1962): 541–542. See also http://embryo.asu.edu/pages/beatrice-mintz, Adam Navis, "Beatrice Mintz," *Embryo Project Encyclopedia.*

16. Nicole Le Douarin and Anne McClaren, eds., *Chimeras in Developmental Biology* (London: Academic Press, 1984). See also http://embryo.asu.edu/pages/nicole-marthe-le-douarin, Brad Peterson, "Nicole Marthe Le Douarin," *Embryo Project Encyclopedia.*
17. Jonathan M. W. Slack, *Stem Cells: A Very Short Introduction* (New York: Oxford University Press, 2012).
18. See, for example, Jane Maienschein, *Whose View of Life? Embryos, Cloning, and Stem Cells* (Cambridge, Mass.: Harvard University Press, 2003). On regulation, see Ronald Michael Green, *The Human Embryo Research Debates: Bioethics in the Vortex of Controversy* (New York: Oxford University Press, 2001).
19. Jane Maienschein, "Stem-Cell Research Utilizing Embryonic Tissue Should Be Conducted," 237–247; Bertha Alvarez Maninnen, "Stem-Cell Research Utilizing Embryo Tissue Should Not Be Conducted," 248–258; and "Joint Reply," 259–260, in *Contemporary Debates in Bioethics,* ed. Arthur L. Caplan and Robert Arp (New York: Wiley, 2013).
20. See "Stem Cell Information: Stem Cells and Diseases," National Institutes of Health, last updated June 05, 2012, http://stemcells.nih.gov/info/pages/health.aspx, and the ClinicalTrials.gov registry website.
21. Steven D Schwartz, Jen-Pierre Hubschman, Gad Heilwell, Valentina Franco-Cardenas, Carolyn K Pan, Rosaleen M Ostrick, Edmund Mickunas, Roger Gay, Irina Klimanskaya, Robert Lanza, "Embryonic Stem Cell Trials for Macular Degeneration: A Preliminary Report," *Lancet* 379 (2012): 713–720.
22. Leroy Stevens, "The Development of Transplantable Teratocarcinomas from Intratesticular Grafts of Pre- and Postimplantation Mouse Embryos," *Developmental Biology* 21 (1970): 364–382.
23. Lisa Landers, "The Unrelenting Headache," *New York Times Magazine,* May 19, 2013, 20, 22, www.nytimes.com/interactive/2013/05/19/magazine/diagnosis-unrelenting-headache.html.
24. One example of current work on hematopoietic, or adult, stem cells brings together heredity, development, and evolution, drawing on understanding of gene regulatory networks. See Ellen Rothenberg's work at the Caltech Division of Biology and Biological Engineering, http://biology.caltech.edu/Members/Rothenberg.

25. Robert Lanza, John Gearhart, Brigid Hogan, Douglas Melton, Roger Pederson, E. Donnall Thomas, James Thomson, and Ian Wilmut, eds., *Essentials of Stem Cell Biology,* 2nd ed. (New York: Academic Press, 2009). For an excellent discussion of stem cell science and its current, planned, and possible applications, as well as limitations, see Christine Mummery, Ian Wilmut, Anja van de Stolpe, and Bernard A. J. Roelen, *Stem Cells: Scientific Facts and Fiction* (London: Academic Press/Elsevier, 2011).
26. K. Takahashi and S. Yamanaka, "Induction of Pluripotent Stem Cells from Mouse Embryonic and Adult Fibroblast Cultures by Defined Factors," *Cell* 126 (2006): 663–676.
27. K. Takahashi, K. Tanabe, M. Ohnuki, M. Narita, T. Ichisaka, K. Tomoda, and S. Yamanaka, "Induction of Pluripotent Stem Cells from Adult Human Fibroblasts by Defined Factors," *Cell* 131 (2007): 861–872; Junying Yu, Maxim A. Vodyanik, Kim Smuga-Otto, Jessica Antosiewicz-Bourget, Jennifer L. Frane, Shulan Tian, Jeff Nie, Gudrun A. Jonsdottir, Victor Ruotti, Ron Stewart, et al., "Induced Pluripotent Stem Cell Lines Derived from Human Somatic Cells," *Science* 318 (2007): 1917–1920.
28. H. Wang, et al. "One-Step Generation of Mice Carrying Mutations in Multiple Genes by CRISPR/Cas-Mediated Genome Engineering," *Cell* 153 (2013): 910–918.
29. For more information on clinical trials, see the ClinicalTrials.gov website: http://clinicaltrials.gov/ct2/results?term=stem+cells&Search=Search.
30. European Human Embryonic Stem Cell Registry, www.hescreg.eu/.
31. Dickey-Wicker Amendment, Pub. L. No. 104-134, 110 Stat. 1321–229 (1996).
32. For discussion, see Kyla Dunn, "The Politics of Stem Cells," *NOVA scienceNOW,* April 1, 2005, www.pbs.org/wgbh/nova/body/stem-cells-politics.html.
33. National Institutes of Health, "National Institutes of Health Guidelines for Research Using Human Pluripotent Stem Cells," 65 FR 51976/65 FR 69951, 2000, http://stemcells.nih.gov/news/newsarchives/pages/stemcellguidelines.aspx.
34. President George W. Bush, "President Discusses Stem Cell Research," The Bush Ranch, August 9, 2001, http://georgewbush-whitehouse.archives.gov/news/releases/2001/08/20010809-2.html.

35. President George W. Bush, Executive Order 13435, "Expanding Approved Stem Cell Lines in Ethically Responsible Ways," June 20, 2007, *Federal Register* 72 (2007), 34591–34593. http://edocket.access.gpo .gov/2007/pdf/07-3112.pdf.
36. President Barack Obama, Executive Order 13505, "Removing Barriers to Responsible Scientific Research Involving Human Stem Cells," March 9, 2009, *Federal Register* 74 (2009): 10667–10668. http://www .gpo.gov/fdsys/pkg/FR-2009-03-11/pdf/E9-5441.pdf.
37. *Dr. James L. Sherley, et al. v. Kathleen Sebelius, et al.*, U.S. District Court for the District of Columbia, Civ. No. 1:09-cv-1575 (RCL) (2010).
38. *Dr. James L. Sherley, et al. v. Kathleen Sebelius, et al.*, Civ. No. 1:09–cv–1575 (RCL), July 27, 2011.
39. Court opinions for *Sherley v. Sebelius*: *Sherley v. Sebelius,* 644 F.3d 388 (2011); *Sherley v. Sebelius,* 704 F.Supp.2d 63 (2010); *Sherley v. Sebelius,* 610 F.3d 69 (2010).
40. Ibid.
41. "Stem Cells: Regenerative Medicine," National Institutes of Health, last updated December 09, 2011, http://stemcells.nih.gov/info/scireport/2006report.htm.
42. Alejandro Sánchez Alvarado, at the University of Utah, www.neuro.utah .edu/people/faculty/sanchez.html and at the Stowers Institute for Medical Research, http://www.stowers.org/faculty/s%C3%A1nchez-lab.

8. Constraints and Opportunities for Construction

1. "29–30 December; Living Systems: Synthesis, Assembly, Origins," short report, *Science* 174 (1971): 858–859.
2. Daniel G. Gibson, John I. Glass, Carole Lartigue, Vladimir N. Noskov, Ray-Yuan Chuang, Mikkel A. Algire, Gwynedd A. Benders, Michael G. Montague, Li Ma, Monzia M. Moodie, et al., "Creation of a Bacterial Cell Controlled by a Chemically Synthesized Genome," *Science* 329 (2010): 52–56. For Venter's thinking, see J. Craig Venter, *Life at the Speed of Light: From the Double Helix to the Dawn of the Digital Age* (New York: Viking, 2013).
3. Presidential Commission for the Study of Bioethical Issues, "New Directions: The Ethics of Synthetic Biology and Emerging Technologies," Washington, D.C., December 2010, http://bioethics.gov/synthetic-biology-report.

4. Ibid., 36. Also see page 111 and following for discussion of ethical principles. The Woodrow Wilson Center maintains a scorecard on how well the government is doing in meeting the recommendations of the Commission; see Woodrow Wilson International Center for Scholars, Synthetic Biology Project, "Synthetic Biology Scorecard," last updated November 29, 2012, www.synbioproject.org/scorecard/.
5. Ewen Callaway, "European Ban on Stem-Cell Patents Has a Silver Lining," *Nature* 478 (2011): 441–442.
6. *Association for Molecular Pathology, et al. v. Myriad Genetics, Inc., et al,* 569 U.S. (2013), no. 12-398.
7. Christine Mummery, Ian Wilmut, Anja van de Stolpe, and Bernard A. J. Roelen, *Stem Cells: Scientific Facts and Fiction* (London: Academic Press/Elsevier, 2011), 143–146. They discuss this inspiring case and include an interview with the engineer who developed the trachea.
8. For an overview of tissue engineering see, for example, Karoly Jakab, Francoise Marga, Cyrille Norotte, Keith Murphy, Gordana Vunjak-Novakovic, and Gabor Forgacs, "Tissue Engineering by Self-Assembly and Bio-printing of Living Cells," *Biofabrication* 2, no. 2 (2010): 022001.
9. As typically occurs in such cases, the first reports appeared in the news media, such as Henry Fountain, "Synthetic Windpipe Is Used to Replace Cancerous One," *New York Times,* January 12, 2012, www.nytimes.com/2012/01/13/health/research/surgeons-transplant-synthetic-trachea-in-baltimore-man.html.
10. Many news reports covered the June 2012 story. For Sasai's account of the research, including previous and subsequent studies, see the website of the RIKEN research facility (www.riken.jp/en/research/labs/cdb/) and its Center for Developmental Biology (www.cdb.riken.jp/en/index.html).
11. Zhenyu Tang, Aijun Wang, Falei Yuan, Zhiqiang Yan, Bo Liu, Julia S. Chu, Jill A. Helms, and Song Li, "Differentiation of Multipotent Vascular Stem Cells Contributes to Vascular Diseases," *Nature Communications* 3 (2012): 875.
12. James J. H. Chong, Vashe Chandrakanthan, Munira Xaymardan, Naisana S. Asli, Joan Li, Ishtiaq Ahmed, Corey Heffernan, Mary K. Menon, Christopher J. Scarlett, Amirsalar Rashidianfar, et al., "Adult Cardiac-Resident MSC-like Stem Cells with a Proepicardial Origin," *Cell Stem Cell* 9 (2011): 527–540.

13. "March of Dimes Awards $250,000 Prize to Two Scientists Who Pioneered Advances in Skin Disorders," press release, March of Dimes, February 29, 2012, www.marchofdimes.com/news/march-of-dimes-awards-250000-prize-to-two-scientists-who-pioneered-advances-in-skin-disorders.aspx.
14. Takanori Takebe, Keisuke Sekine, Masahiro Enomura, Hiroyuki Koike, Masaki Kimura, Takunori Ogaeri, Ran-Ran Zhang, Yasuharu Ueno, Yun-Weng Zheng, Naoto Koike, et al., "Vascularized and Functional Human Liver from an iPSC-derived Organ Bud Transplant," *Nature* 499 (2013): 481–484.
15. F. X. Jiang FX and G. Morahan, "Pancreatic Stem Cells: From Possible to Probable," *Stem Cell Review* 3 (2012): 647–657.
16. Dongeun Huh, Benjamin D. Matthews, Akiko Mammoto, Martín Montoya-Zavala, Hong Yuan Hsin, and Donald E. Ingber, "Reconstituting Organ-Level Lung Functions on a Chip," *Science* 328 (2010): 1662–1668. Also see Elizabeth Dougherty, "Living, Breathing Human Lung-on-a-Chip," Harvard Gazette, June 24, 2010, http://news.harvard.edu/gazette/story/2010/06/living-breathing-human-lung-on-a-chip/, and Wyss Institute at Harvard, "Lung on a Chip," updated January 29, 2013, http://wyss.harvard.edu/viewpage/240/, a marvelous award-winning video demonstration.
17. Sam Kean, "Blogging the Human Genome," *Slate,* July 11, 2012, www.slate.com/articles/health_and_science/chromosomes/features/2012/blogging_the_human_genome_/blogging_the_human_genome_craig_venter_and_the_race_to_sequence_the_human_genome_.html; Jill Adams, "Sequencing Human Genome: The Contributions of Francis Collins and Craig Venter," *Nature Education* 1 (2008): 1. www.nature.com/scitable/topicpage/sequencing-human-genome-the-contributions-of-francis-686.
18. Gibson et al., "Creation of a Bacterial Cell"; Presidential Commission, "New Directions."
19. Jonathan R. Karr, Jayodita C. Sanghvi, Derek N. Macklin, Miriam V. Gutschow, Jared M. Jacobs, Benjamin Bolival, Nacyra Assad-Garcia, John I. Glass, and Markus W. Covert, "A Whole-Cell Computational Model Predicts Phenotype from Genotype," *Cell* 150 (2012): 389–401.
20. Elizabeth Penisi, "How Do Microbes Shape Animal Development?," *Science* 340 (2013): 1159–1160.
21. Leroy Hood, "Tackling the Microbiome," *Science* 336 (2012): 1209.

Therefore . . .

1. President George W. Bush, "President Discusses Stem Cell Research," The Bush Ranch, August 9, 2001, http://georgewbush-whitehouse.archives.gov/news/releases/2001/08/20010809-2.html.
2. Sanctity of Human Life Act, H.R. 212, 112th Cong. (2011), http://thomas.loc.gov/cgi-bin/query/z?c112:H.R.212.
3. "Personhood Protects Human Dignity," www.personhood.net/index.php?option=com_content&view=article&id=267:cloning-always-takes-a-human-life&catid=129:state-constitutional-amendments&Itemid=567.
4. See http:personhoodusa.com (accessed August 21, 2013).
5. Susan J. Lee, Henry J. Peter Ralston, Eleanor A. Drey, John Colin Partridge, and Mark A. Rosen, "Fetal Pain: A Systematic Multidisciplinary Review of the Evidence," *Journal of the American Medical Association* 294 (2005): 947–954.
6. See, for example, this early summary: Curtis L. Lowery, Mary Hardman, Nirvana Manning, R. Whit Hall, K. J. S. Anand, and Barbara Clancy, "Neurodevelopmental Changes of Fetal Pain," *Seminars in Perinatology* 5 (2007): 275–282.
7. A review of scholarly articles on "fetal pain," available with a key word search at the National Library of Medicine's PubMed service (http://www.ncbi.nlm.nih.gov/pubmed) provides an excellent, up-to-the minute collection of research results.
8. Erik Eckholm, "Theory on Pain Is Driving Rules for Abortions," *New York Times,* August 1, 2013, www.nytimes.com/2013/08/02/us/theory-on-pain-is-driving-rules-for-abortions.html.

Acknowledgments

My work on understanding embryos began in school, indirectly inspired by some very dedicated teachers. My high school in Oak Ridge, Tennessee, taught something called biology. But in the late 1960s, neither reproductive topics nor evolution was considered a legitimate subject for study in public schools. Under the guidance of a very wise and wonderful high school English teacher, we read Darwin's *On the Origin of Species* as literature. We had to read it there because until the 1967–1968 school year it was still against Tennessee's "Scopes Law" to teach evolution as biology in the schools. The only discussion of reproduction came in Girl Scouts, where the leaders insisted that young women should understand how sex and development work—or else bad things could happen.

College studies at Yale University with John P. Trinkaus (known as Trink) and Frederic Lawrence Holmes (known as Larry) gave me an introduction to developmental biology and to the history of science, which led to the exciting realization that developmental biology has a fascinating history. Graduate school took me to Indiana University in Bloomington to work with Fred Churchill in history and philosophy of science. Churchill had attracted a wonderful group of graduate students including John Beatty, Ronald Rainger, and others who became life-long friends and colleagues. At Indiana, I also had the privilege of working with embryologist Robert Briggs in class and as a member of my dissertation committee. In addition to my making life-long friends and collaborators, two inspiring things happened during that time: my dissertation study with Briggs and my research visit to the Marine Biological Laboratory (MBL) in Woods Hole, Massachusetts.

To learn more about developmental biology, I took Briggs's embryology course. There, he cheerfully insisted that we must learn for ourselves to develop the tools and techniques for embryo experiments. By that time in the 1970s, we could easily have ordered equipment from a supply company, but he wanted us to experience the challenge of successfully creating our own glass needles in particular. He taught us to put a small rod of glass over the Bunsen burner, then to pull out the softened glass at just the right time and for just the right distance. The consistency is much like taffy, as it melts and then quickly congeals as it cools. What becomes clear when you do this for yourself (which you should not do without proper equipment, of course) is that you gain an incredibly sharp, precise tool if you get it just right. You also learn how clunky the results can be when you (usually) do not. Finding or making the appropriate tools to perform the job is so very important in science.

I then applied for a dissertation improvement grant from the National Science Foundation. These small grants have provided opportunities for hundreds of thousands of young researchers to expand their horizons and discover things they could never have learned without the support. They make graduate student research possible. My award, made by program officer Ronald Overmann in the History and Philosophy of Science Program, had a considerable impact on my direction of research because it took me to the MBL in late May and June when a number of marine species were breeding most actively. I wanted to reproduce the observations and experiments of the leading embryologists around 1900, when the research carried out at the MBL and the Stazione Zoologica in Naples was just beginning to reveal how embryos become organized individual organisms.

The MBL very generously let me sit in a laboratory in the Lillie Building and loaned me the oldest equipment around to mimic the good old days when biologists used natural light from the high windows to see through the excellent microscopes. I went down to the docks and to the Marine Resources Center to collect embryos and watch what other people were doing. I read a lot and listened a lot. Then some of the older guys (all guys, as it turned out—there were women scientists around as well, but few in the older generations) started to drop by my "lab" and kibitz. They would look at my preparations and say, "That's not the way Wilson (or Morgan or Harrison) would have done it." Then "Here's what you need to do." There is no better way to learn than to watch an expert. I learned to see through the microscope in a way I never had before.

When I tried to reproduce the observations that Wilson had made on the worm *Nereis,* a senior embryologist from the University of North Carolina who had studied these worms himself, Donald Costello, helped me. He told stories about his own youthful days when he would go down to the dock, lie on his stomach, and scoop up the worms that were spiraling in an upward mating dance toward the light of the full moon (or flashlight). He made clear that if I did the same thing, I would understand what those embryologists a century before had been doing and why, and I would also begin to find new questions myself. He was right.

How could one not fall in love with the passion for discovery at the MBL? In fact, I have written almost all of this book while sitting in a lovely office overlooking the water in Woods Hole. A cool breeze blows through, and (rather noisy) birds sing a happy accompaniment, with one of the world's most well-loved libraries for the history of biology nearby to provide just about anything anybody could ever need for such research. I am not far from where most of the leading embryologists of the twentieth century studied and worked, at least for some part of their careers. For over 125 years, the MBL has been a mecca for students, who can study embryology or take other, more specialized courses. There is so much to learn.

Does it matter to have such opportunities? Of course it does, because actually experiencing nature and watching embryos develop provide a concrete picture of what is happening and make it difficult to hold on to socially constructed imaginations that do not actually fit with this observed reality. Fortunately, the Internet provides access to many videos and other ways to experience embryonic development even if one does not have a laboratory handy. The Society for Developmental Biology, American Society for Cell Biology, the Embryo Project Encyclopedia, and the History of MBL Project all offer rich resources for those eager to discover and to embrace biological knowledge. The MBL Community Archives Project, inspired by Matt Person, expands that knowledge by seeking to place biological discoveries in the context of the scientific and family community. Thanks to all the previous embryologists and all my teachers for making this world come alive. Thanks also to the NSF for a sequence of grants that have made my research possible.

These NSF grants, supported especially by permanent program officers for the Science, Technology, and Society Program (in its various incarnations)

Ronald Overmann and Fred Kronz, have made possible the Embryo Project. This project in turn has led to the Embryo Project Encyclopedia and training program. A group of postdocs, graduate students, and colleagues have been joined by a cycle of undergraduates who all work together on our Embryo Project Encyclopedia (see http://embryo.asu.edu and http://history.archives.mbl.edu, for example).

In addition to publishing books and articles, we are developing open-access research materials and systems to link our scholarship to the work of others. See http://hpsrepository.asu.edu for more on this project and its participants. In particular, thanks to our wonderful Center Program Manager and collaborator Jessica Ranney, to the editorial team of Steve Elliott, Kate MacCord, Erica O'Neil, Erick Peirson, Julia Damerow, Valerie Racine, and Florian Huber, and to postdocs Mary Sunderland, Grant Yamashita, and Nathan Crowe for helping that project take great leaps forward. Stephanie Crowe, Michael Dietrich, Cathy Norton, Diane Rielinger, Matt Person, Kristen Lans, and Nathan Wilson have helped to establish a digital archive for History and Philosophy of Science Projects (http://digitalhps.asu.edu) in ways that take what we have learned from the Embryo Project and apply them to other projects as well.

Thanks also to Michael Fisher at Harvard University Press, who continues to support work that does not fall into the normal categories and to have confidence in his authors. Thanks to our Arizona State University and Marine Biological Laboratory leaders, especially Michael Crow and Gary Borisy, for providing so much support to my wonderful colleague Manfred Laubichler and me. They made possible the historical and philosophical (HPS) efforts of the collaborative ASU-MBL HPS Program from 2011 to 2013, and the continuing history projects and digital presence thereafter.

Catherine May, Jason Robert, Manfred Laubichler, and Lijing Jiang read chapters or the whole and offered valuable suggestions. Catherine May also offered ideas about illustrations and provided the embryo images used on the cover. Project manager and graduate student Kate MacCord has read the entire manuscript more than once. At one point, she told me that my ideas were muddled, then she proceeded to suggest exactly what I needed to set them right. Kate is brilliant at knowing how to help people do better work. And thanks as always to Rick Creath for reading everything and then patiently helping to make it better by explaining what I really meant. He also provided just the right number of his most amazing pastries and tasty cooking at just the right time. Producing a book really is a team effort. Thanks to the team.

Index